Anatomical Variations in Clinical Dentistry

Joe Iwanaga • R. Shane Tubbs
Editors

Anatomical Variations in Clinical Dentistry

Springer

Editors
Joe Iwanaga
Seattle Science Foundation
Seattle, WA
USA

R. Shane Tubbs
Seattle Science Foundation
Seattle, WA
USA

ISBN 978-3-319-97960-1 ISBN 978-3-319-97961-8 (eBook)
https://doi.org/10.1007/978-3-319-97961-8

Library of Congress Control Number: 2018962487

This Springer imprint is published by the registered company Springer Nature Switzerland AG
The registered company address is: Gewerbestrasse 11, 6330 Cham, Switzerland

Preface

Anatomical variations are encountered on a daily basis by those specialists entering the oral cavity (e.g., dentists and oral surgeons) and adjacent regions. Therefore, for optimal daily clinical practice, both trainees and professionals in these fields and others (e.g., endodontists, periodontists, implantologists, anatomists, maxillofacial surgeons, otolaryngologists, dental students, and dental hygienists) should be aware of the most common variants found in the oral cavity. *Anatomical Variations in Clinical Dentistry* seeks to provide a go-to reference on this topic. The book begins by introducing the reader to anatomical variations from the point of view of different clinical practitioners—oral and maxillofacial surgeons, periodontists, and endodontists. The newest anatomical knowledge and variations are then presented in turn for the mandible, maxillary sinus, hard palate, floor of the mouth, lips, temporomandibular joint, and teeth. In each chapter, clinical annotations are included in order to enhance the understanding of the relationships between surgery and anatomy. The internationally renowned authors of the text have been carefully selected for their expertise.

Seattle, WA, USA — Joe Iwanaga
Seattle, WA, USA — R. Shane Tubbs

Contents

Part I

Anatomical Variations from the Point of view of Clinical Practitioners

Anatomical Variations Relevant to Oral and Maxillofacial Surgeons

1

Jingo Kusukawa

Anatomy provides the foundation of surgery and is the most basic and essential science in surgery. It is indispensable not only for meeting diagnostic challenges but also for developing surgical procedures. Therefore, anatomy is a basic requirement for all surgical specialties.

Oral and maxillofacial surgery specializes in treating pathological conditions and disorders, injuries, defects, deformities, and malformations in the hard and soft tissues of the oral cavity, jaws, face, and adjacent organs. Therefore, oral and maxillofacial surgeons require specific skills that also necessitate detailed knowledge of the relationships among the anatomical structures encountered during surgical operations in this area. Insufficient anatomical knowledge can result in serious complications and poor cosmetic or functional postoperative results. Clinical anatomy, giving consideration to practical surgery, is crucial for enhancing diagnostic efficiency and performing a safe and effective operation. In addition to fundamental knowledge of the clinical anatomy of the oral and maxillofacial region, we should deepen our understanding of anatomical changes during aging and anatomical variations among individuals.

Morphological changes caused by disuse atrophy of alveolar and jaw bones following tooth loss handicap the surgeon's search for anatomical landmarks such as the piriform aperture and anterior nasal spine, incisive papilla (incisive canal), hamular notch, and neural foramina. Decrease of the alveolar ridge height increases the risk for injury to nerves and blood vessels emerging from neural foramina including the greater palatine foramen (Fig. 1.1), the infraorbital foramen, and the mental foramen. Narrowing of the alveolar ridge increases the risk that the surgeon's scalpel blade leaves the alveolar crest and cuts deeply inside.

Furthermore, anatomical variations complicate the situation of the operation. Such variations are potential risk factors in oral and maxillofacial surgery. In

J. Kusukawa (✉)
Dental and Oral Medical Center, Kurume University School of Medicine, Kurume, Japan
e-mail: kusukawa@med.kurume-u.ac.jp

J. Iwanaga, R. S. Tubbs (eds.), *Anatomical Variations in Clinical Dentistry*,
https://doi.org/10.1007/978-3-319-97961-8_1

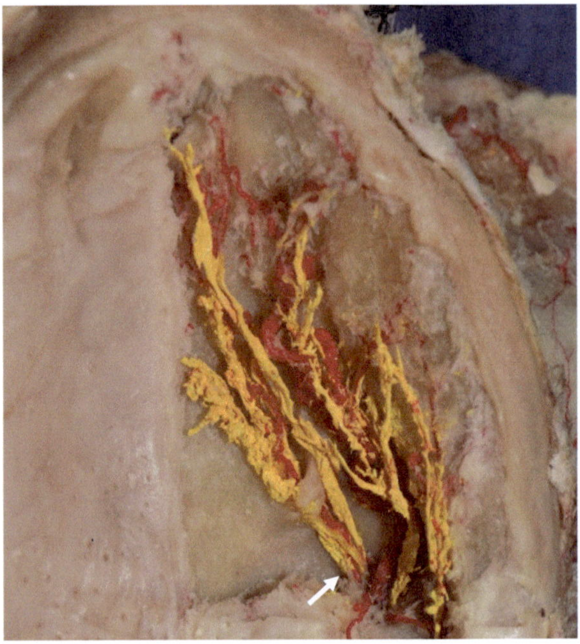

Fig. 1.1 Greater palatine foramen (arrow), nerve, and artery

particular, variations of nerves and blood vessels entail the risk of serious complications. Uncontrollable bleeding from blood vessel injuries can be life-threatening. Permanent nerve injury has a serious negative effect on the patient's quality of life. To avoid such complications, surgeons should be familiar with anatomical variations as well as preventive care.

Third molar surgery remains the commonest procedure carried out by general dentists and oral surgeons. One of the main concerns in third molar surgery is inferior alveolar nerve (IAN) injury. The incidence of permanent IAN injury ranges from 0.35% to 8.4% (Sarikov and Juodzbalys 2014). Apart from third molar surgery, the IAN is at risk of trauma during oral surgery operations such as jaw cyst surgery, or implant surgery. The IAN enters the mandible from the mandibular foramen, which is inside the mandibular ramus, passes through the mandibular canal, leaves from the mental foramen outside the mandibular body, and is distributed over the lower lip and mental region as the mental nerve (MN). To avoid unnecessary complications, we should recognize that there are many variations in the pathway of the IAN and MN. It is well known that the positional relationship between the third molar and mandibular canal corresponds to the incidence of IAN injury. In addition, the prevalence of accessory mental foramina ranges from 2.0% to 13.0% (Iwanaga et al. 2015). Such accessory foramina in the vicinity of the mental foramen are a potential risk for paresthesia of the lower lip and mental region.

Lingual nerve (LN) paralysis is serious complication in oral surgery. As the LN runs along the lingual aspect of the mandible and reaches the tongue through the floor of the mouth, it can be injured by oral surgery operations including third molar surgery, jaw resection, grafting of the alveolar crest, salivary gland surgery,

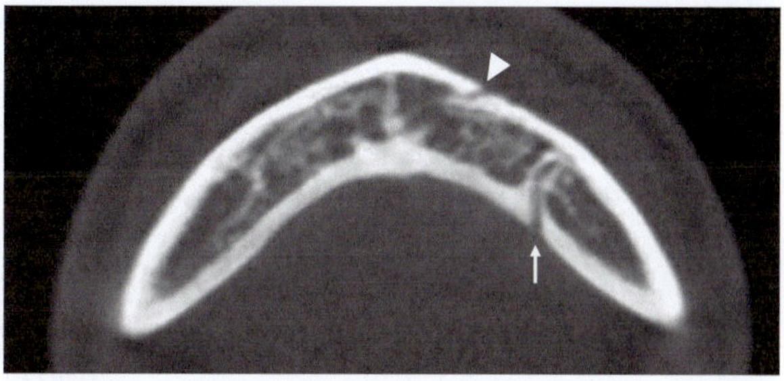

Fig. 1.2 Mandibular lingual canal in the canine-premolar region (arrow). Additional buccal bony canal in the incisal region (arrowhead)

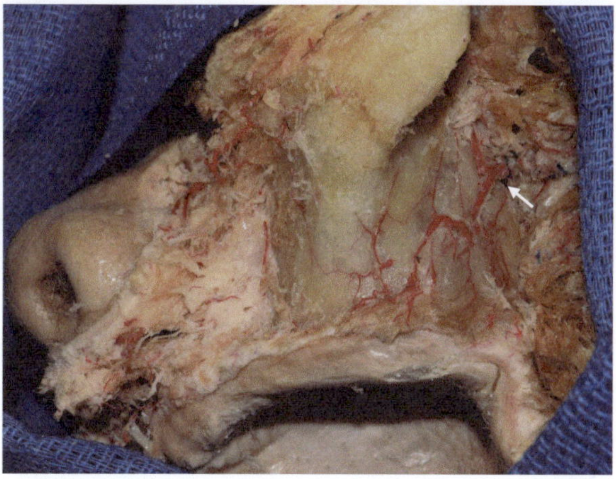

Fig. 1.3 Left posterior superior alveolar artery (arrow)

placement of implants, and tumor excision. The incidence of such complications ranges from 0.04% to 22% (Dias et al. 2015). Although nerve injury is a rare complication, its implications can be significant for the patient.

The sublingual artery (SLA) is a major branch of the lingual artery, branches of which supply the floor of the mouth and occasionally enter the mandible through lingual bony canals (Tepper et al. 2001; Sahman et al. 2014) (Fig. 1.2). During intraosseous surgery of the mandible such as implant placement, the lingual bony canals present a risk for SLA injury, resulting in the development of a large hematoma in the floor of the mouth, which can lead to life-threatening airway obstruction. The posterior superior alveolar artery (PSAA) (Fig. 1.3), one of the terminal branches of maxillary artery, supplies blood to the lateral wall of the maxillary sinus and the sinus floor membrane (Maridati et al. 2014). Injury to the PSAA in sinus-related surgery represented by sinus floor augmentation can cause massive perioperative bleeding (Güncü et al. 2011). To avoid traumatizing these blood vessels, locating their exact position prior to the operation is mandatory.

To perform a safe and reliable surgical operation without complications, surgeons need to be aware of the anatomical variations and detect them precisely prior to surgery, in addition to understanding normal structures. Preoperative detection of anatomical variations of soft tissues is limited. Although angiography, angiographic CT and magnetic resonance angiography (MRA) are the main modalities for detecting the vascular distribution and its variations, they are highly invasive and costly. Additionally it is extremely difficult to detect the nerves in the soft tissue. Operation under direct vision is the most reliable method for detecting and handling them. Anatomical landmarks are also useful for ensuring that the surgical operation proceeds safely and smoothly.

On the other hand, radiographic imaging can identify anatomical variations in bone. The orthopantomogram (OPG) has classically been used to evaluate the jaws, including the maxillary sinus, the mandibular canal and the mental foramen, and their relationships to the teeth. However, OPG does not permit three-dimensional (3D) assessment of their position or visualization of the lingual canals. Multidetector CT and cone beam CT (CBCT) permit finely detailed visualization of the osseous architecture with high resolution and accuracy. Multiplanar reconstruction (MPR) imaging provides surgeons with more detailed anatomical information for recreating two-dimensional (2D) images in optionally different planes and 3D volumetric views. MPR imaging, as provided by MDCT/CBCT, is undoubtedly increasing diagnostic efficiency and surgical accuracy, but it is a novelty for dentists and oral and maxillofacial surgeons. In addition to the development of high-resolution imaging equipment, digital technology has been integrated into surgical planning and procedures. In oral and maxillofacial surgery, computer-aided design/computer-aided manufacturing (CAD/CAM) techniques have been developed to build surgical guides and mock-up 3D full-size models in an attempt to improve the precision of surgical operations such as implant placement, tumor resection, and reconstruction of the jaw. In the near future, robotic surgery and artificial intelligence will be applied to clinical oral and maxillofacial surgery. Such innovations in oral and maxillofacial surgery will improve the safety and accuracy of operations drastically. Whatever the progress in diagnostic and surgical technology, we should remember that anatomy is fundamental and essential for surgeons to secure and make advances in oral and maxillofacial surgery.

References

Dias GJ, de Silva RK, Shah T et al (2015) Multivariate assessment of site of lingual nerve. Br J Oral Maxillofac Surg 53:347–351

Güncü GN, Yildirim YD, Wang HL et al (2011) Location of posterior superior alveolar artery and evaluation of maxillary sinus anatomy with computerized tomography: a clinical study. Clin Oral Implants Res 22:1164–1167

Iwanaga J, Saga T, Tabira Y et al (2015) The clinical anatomy of accessory mental nerves and foramina. Clin Anat 28:848. https://doi.org/10.1002/ca.22597

Maridati P, Stoffella E, Speroni S et al (2014) Alveolar Antral artery isolation during sinus lift procedure with the double window technique. Open Dent J 8:95–103

Sahman H, Ercan A, Elif S et al (2014) Lateral lingual vascular canals of the mandible: a CBCT study of 500 cases. Surg Radiol Anat 36:865–870

Sarikov R, Juodzbalys G (2014) Inferior alveolar nerve injury after mandibular third molar extraction: a literature review. J Oral Maxillofac Res 29:5. https://doi.org/10.5037/jomr.2014.5401

Tepper G, Hofschneider UB, Gahleitner A et al (2001) Computed tomographic diagnosis and localization of bone canals in the mandibular interforaminal region for prevention of bleeding complications during implant surgery. Int J Oral Maxillofac Implants 16:68–72

Anatomical Variations from the Point of View of the Periodontist

2

Daniel E. Shin

2.1 Introduction

Periodontal surgery encompasses a wide range of surgical therapies that are designed to restore and regenerate the natural form and function to lost and damaged structures of the teeth. Common types of periodontal surgical therapies include, but are not limited to, open flap debridement, osseous resective surgery, guided tissue regeneration (GTR), and soft tissue augmentation. It is important for a periodontist to have sound clinical judgment and proficient skills to perform the aforementioned surgical procedures, but it is equally important for the periodontist to have a keen anatomical mind with the means to appreciate the subtle nuances or the overt and pronounced anatomical variations that lie in the vicinity of the periodontal surgical field. Furthermore, it is important for the periodontist to anticipate these anatomical variations and the role they may play in determining the scope of the planned surgical procedure. Thus, in order to safely and properly execute the surgical procedure, the periodontist must have a strong understanding of the precise location of anatomical structures but also be aware of variations with respect to size, shape, and location of vital oral anatomical structures and landmarks. The focus of this chapter is to highlight common variations in anatomical oral structures, which are regularly encountered in periodontal surgery and implant surgery. The reader is referred to subsequent chapters in this textbook for a more in-depth description of these anatomical structures and landmarks.

D. E. Shin (✉)
Department of Periodontics, Indiana University School of Dentistry, Indianapolis, IN, USA
e-mail: danshin@iu.edu

J. Iwanaga, R. S. Tubbs (eds.), *Anatomical Variations in Clinical Dentistry*,
https://doi.org/10.1007/978-3-319-97961-8_2

2.2 Common Anatomical Variations Relevant to Periodontal Surgery

The maxillary palate is an area which requires caution and care due to variation of the precise location of the greater palatine neurovascular bundle and its associated foramen. The extent and depth of the surgical incision and the type of flap elevation are dictated by the location of the greater palatine foramen and the position of the greater palatine neurovascular bundle.

2.2.1 Greater Palatine Foramen and Greater Palatine Neurovascular Bundle

Although the incidence is low, any iatrogenic injury of either the greater palatine foramen or its neurovascular structure may result in bleeding and/or paresthesia (Harris et al. 2005). Thus, the surgeon must have a clear understanding of the precise location and possible anatomical variations associated with these aforementioned structures. The greater palatine neurovascular bundle descends through the greater palatine canal and emerges out of an opening in the hard palate known as the greater palatine foramen. Generally, the greater palatine foramen is approximately 3–4 mm anterior to the posterior border of the hard palate and 15 mm from the palatal midline. From there, the greater palatine neurovascular bundle courses anteriorly in a groove that is formed at the junction of the palatine and alveolar processes.

2.2.1.1 Variations in the Anatomic Configuration of the Greater Palatine Foramen

A clear understanding of the location and anatomy of the greater palatine foramen is needed prior to performing any type of periodontal surgery in the maxillary palate. However, the precise location and the position of the greater palatine foramen can vary among individuals. A systematic review and meta-analysis identified five possible positions for the greater palatine foramen: (1) anterior to the mesial surface of the second molar, (2) in between the mesial surface and the distal surface of the second molar, (3) at the interproximal level between the second and third molars, (4) in between the mesial surface and distal surface of the third molar, and (5) distal to the third molar (Tomaszewska et al. 2014). Out of these five possible locations, the greater palatine foramen was most often located in between the mesial surface and the distal surface of the third molar—on average, this was seen in two-thirds to three-quarters of all patients, both in Europe and worldwide. On the other hand, in another study that examined the location of the greater palatine foramen from skull dissections, it was found that the main location of the greater palatine foramen was directly palatal to the second molar (Klosek and Rungruang 2009). In another skull dissection study, Fu et al. (2011) found that the most common location of the greater palatine foramen was between the second and third molar teeth. Such uncertainty in the precise anatomical location of the greater palatine foramen makes the maxillary posterior palatal region a challenge to perform a wide variety of anesthetic, dental, and surgical procedures.

2.2.1.2 Variations in the Anatomic Configuration of the Greater Palatine Neurovascular Bundle

After exiting the greater palatine foramen, the greater palatine neurovascular bundle courses through a groove that is formed at the junction of the palatine and alveolar processes. For a periodontist, it is important to know the location of the greater palatine neurovascular bundle because dissection of the palatal mucosa can be dictated by the shape of the palatal vault. In general, the location of the greater palatine bundle from the CEJ is correlated to the height of the palatal vault: for a low palatal vault, the average distance between the greater palatine artery and the CEJs of maxillary molar teeth is 7 mm; for an average palatal vault, the mean distance is 12 mm; and for a high U-shaped palatal vault, the average distance is 17 mm (Reiser et al. 1996).

For a periodontist, the greater palatine neurovascular bundle has two important clinical significances. First, when performing palatal surgery, vertical incisions in the posterior palate should be avoided to eliminate the risk of iatrogenically injuring the greater palatine neurovascular bundle. Second, when harvesting autogenous subepithelial connective tissue grafts, the clinician must be aware that harvesting the palatal graft tissue more than 7 mm from the CEJs of maxillary molars in a shallow vault, 12 mm in an average vault, and 17 mm in a high palatal vault may lead to undesirable complications, such as hemorrhaging or nerve damage.

2.2.2 Anatomy and Variations of the Maxillary Sinus

Structurally, when one thinks of the maxillary sinus, the image of a house comes to mind. Externally, the antrum (cavity) of the maxillary sinus is surrounded by six bony walls on the anterior, medial, posterior, lateral, superior, and inferior sides. Internally, the walls of the aforementioned sinuses are lined by a mucoperiosteal lining known as the Schneiderian membrane. Understanding the anatomy and the relevance of these bony walls and the mucoperiosteal lining is crucial for the periodontist due to the close association of blood vessels and nerves to the sinus walls.

The anterior wall of the maxillary sinus contains the infraorbital nerve and a branch of the infraorbital artery known as the anterior superior alveolar artery. The anterior superior alveolar artery feeds blood to the anterior portion of the mucous membrane of the maxillary sinus and to the anterior teeth (from incisor to canine) and its surrounding periodontium. When preparing the osteotomy (access) via a lateral approach, the access preparation line should be made at a safe distance from the anterior wall to avoid disrupting branches of the infraorbital nerve and blood vessels to the maxillary anterior teeth.

The medial wall of the sinus separates the maxillary sinus from the neighboring nasal cavity. The superior portion of the medial wall of the sinus contains the ostium, which is an opening that drains mucous from the maxillary sinus into the middle meatus of the nasal cavity. It is vital to keep this ostium patent since the health of the sinus is dependent on proper drainage of both mucous and bacteria out of the maxillary sinus. As such, when performing a sinus lift procedure, the periodontist must be careful to avoid overfilling the sinus with graft material which could potentially occlude the ostium. If this were to occur, sinusitis may occur.

The posterior wall of the maxillary sinus separates the antrum from the pterygopalatine fossa. Within the posterior wall reside the posterior superior alveolar blood vessels and nerves, which supply the maxillary molars. Of greater concern is that the posterior wall also contains the pterygoid plexus of veins, which communicates with the cavernous sinus. Potentially, any infection from the maxillary sinus region or any surrounding areas can travel through the pterygoid venous plexus, anastomose superiorly with the cavernous sinus via the emissary veins, and spread to the cavernous sinus. This might result in cavernous sinus thrombosis which is a life-threatening condition and requires immediate medical attention.

The lateral wall of the maxillary sinus receives its blood supply from the posterior superior alveolar artery and the infraorbital artery. This presents a challenge for the periodontist preparing an osteotomy via a lateral approach because any inadvertent iatrogenic injury to either the posterior superior alveolar artery or the infraorbital artery during the surgical procedure may result in any one of the following unnecessary surgical complications, such as excessive hemorrhaging, poor visualization of the surgical field, and interfering with the fixation of the graft material. As a result, a presurgical cone beam computed tomography (CBCT) assessment of the sinus walls may be warranted to determine the precise location of the posterior superior alveolar artery within the lateral wall of the sinus.

The superior wall of the maxillary sinus corresponds to the orbital floor. It is a thin and fragile plate of bone that must be treated delicately when performing sinus lift procedures.

The inferior wall of the maxillary sinus corresponds to the floor of the sinus, and it may extend between the roots of the maxillary molars.

Within the maxillary sinus, a thin mucoperiosteum known as the Schneiderian membrane lines the bony walls of the sinus cavity. Histologically, the Schneiderian membrane is a bilaminar layer with an inner portion (that faces the antrum of the sinus) and an outer portion (that faces the bony wall.) The inner portion of the Schneiderian membrane is lined by ciliated columnar epithelium which has the crucial function of "sweeping" bacteria, debris, and toxins out of the maxillary sinus and toward the ostium. For a periodontist, it is important to keep in mind that the Schneiderian membrane is a thin lining that ranges from 0.3 to 0.8 mm in healthy individuals with no maxillary sinus pathology (Mogensen and Tos 1977). Consequently, perforations of the sinus membrane are a common complication that occurs during sinus elevation procedures. However, in individuals suffering from chronic sinusitis, the Schneiderian membrane may be hyperplastic and thicker than normal. In such a case, it is recommended that the patient be evaluated by an ENT prior to any sinus elevation surgery.

Emerging from the surrounding bony antral walls are strut-like bony cortical projections known as maxillary sinus septa. These projections were first described by Underwood in 1910. At that time, the maxillary sinus septum did not pose much of a clinical significance. However, with the more recent advent of pre-implant sinus elevation procedures, a maxillary sinus septum poses serious challenges to the periodontist due to the wide variation in their precise location, prevalence, and etiology. According to Underwood—who first discovered and described the maxillary sinus septum—the septum is most commonly found in the posterior region of the sinus (Underwood 1910). However, in a more recent study, Krennmair et al. (1999) found that 75% of septa were located in the anterior/premolar region, while Kim et al. (2006) found that the majority of septum was located in the middle region (distal aspect of the second premolar to the distal aspect of the second molar). Moreover, the prevalence of a maxillary septum varies from 16% to 58% according to the various studies (Krennmair et al. 1999 ; Underwood 1910). From a periodontal surgical standpoint, knowing if a septum is present, and where its precise location is, is imperative. Osteotomy procedures via the lateral approach require the fracturing of the lateral antral wall in a superior and medial direction (lateral hinge door formation). If a sinus septum is present on the septal floor or any other location, the sinus membrane may be inadvertently perforated upon elevation. Various modifications have been proposed to avoid the tearing of the Schneiderian membrane when a septum is present. Boyne and James (1980) recommended cutting and removing the bony septum so that bone graft can be placed over the entire antral floor without interruption. Tidwell et al. (1992) recommended subdividing the lateral antral wall into the anterior and posterior parts and then inverted both trap doors so as to allow elevation of the membrane without tearing it. Whichever modified technique is used, however, the periodontist must have accurate information and a clear understanding of the patient's maxillary sinus to avoid intraoperative and/or postoperative complications.

2.3 Mandibular Anatomical Structures Relevant to Periodontal Surgery

Several important anatomical structures in the mandible are regularly encountered in periodontal surgery. As such, they deserve much discussion due to the great surgical importance they have in periodontal and implant surgery.

2.3.1 Mandibular Canal

The mandibular canal houses the inferior alveolar nerve and blood vessels. It begins at the mandibular foramen on the medial surface of the mandibular ramus and courses inferiorly and anteriorly until it becomes horizontal apical to the root tip of the molars. As the mandibular canal goes forward, the distance between the apices in the molars and the canal increases.

2.3.1.1 Variations in the Anatomic Configuration of the Mandibular Canal

On a two-dimensional conventional intraoral radiograph, the mandibular canal typically appears to gracefully descend as it proceeds anteriorly toward the mental foramen. However, the mandibular canal may actually have different types of anatomic configurations that belie its typical appearance on conventional radiographs. For instance, in a review conducted by Anderson et al. (1991) of the intraosseous course of the inferior alveolar nerve, the buccolingual and the superior-inferior positions of the mandibular canal rarely showed consistencies among different mandibles. Rather, the mandibular canal was shown to either have a sharp decline or it followed a concave, tortuous course with a posterior segment descending as it progressed anteriorly and an anterior segment that ascended to the mental foramen. Moreover, in approximately 1% of cases, the mandibular canal bifurcates in the superior-inferior plane or the medial-lateral plane within the body of the mandible, thereby giving off two different canals and two different mental foramens which may not be discernable from a conventional radiograph (Dario 2002). Consequently, because of its different anatomic configurations, any procedure that brings the surgery close to the inferior alveolar nerve would be risky without obtaining a presurgical CT scan of the area.

2.3.2 Mental Nerve and Foramen

The mental foramen is the exit point of the terminal branches of the inferior alveolar nerve and associated vessels that supply the lower lip and the chin.

2.3.2.1 Variations in the Anatomic Configuration of the Mental Nerve and Foramen

The mental foramen has a high degree of variability with regard to its position and location in both the horizontal and vertical planes. With respect to its variations in the horizontal plane, several studies show that this variation can be related to race: in Caucasians, the mental foramen was seen in between the apices of the premolars (Fishel et al. 1976), while in Chinese individuals, the mental foramen was observed to be next to the apex of the second premolar (Wang et al. 1986). In another study, the mental foramen was found to be in closer proximity to the canine or molar teeth (Ngeow and Yuzawati 2003). Likewise, the anatomical position and the location of the mental foramen can vary in the vertical plane with respect to the apices of both the first and second premolar. In a report evaluating the vertical position of the mental foramen in relation to the apices of the first premolar teeth, it was found that the foramen was coronal to the apex in approximately 38.6% of cases, at the apex in 15.4% of cases, and apical to the apex in 46.0% of cases (Fishel et al. 1976). In the same report, it was found that the foramen's location, in relation to the *second* premolar, was coronal to the apex in 24.5% of cases, at the apex in 13.9% of cases, and apical to the apex in 61.6% of cases. Based off of this report, it can be concluded that the periodontist must exercise extreme caution when placing immediate implants in the premolar area, because in 25–38% of cases the foramen is located coronal to the premolar's apex.

Moreover, periodontal surgery can also be complicated by the unknown course and extent of the inferior alveolar nerve as it proceeds into the direction of the mental foramen. As the inferior alveolar nerve makes its way to the mental foramen, it can run anterior to the mental foramen and then courses upward, outward, and backward to eventually exit out of the foramen. This is known as the "anterior loop" (Greenstein et al. 2008). While the anterior loop has a high prevalence, panoramic and periapical radiographs are not accurate assessment tools to detect the presence of the anterior loop due to the high false-positive and false-negative rates (Greenstein et al. 2008). Thus, as a general rule, when performing periodontal implant surgery in the mandibular premolar region, a CBCT scan should be obtained prior to surgery to ascertain the precise location of the anterior loop and the mental foramen.

2.3.3 Incisive Branch of the Inferior Alveolar Nerve

The incisive branch of the inferior alveolar nerve—as it continues medially to the anterior mandibular region, provides innervation to the anterior mandibular teeth and canines.

2.3.3.1 Variations in the Anatomic Configuration of the Incisive Branch of the Inferior Alveolar Nerve

The incisive branch courses through the intramedullary spaces and may not be evident on conventional radiographs and is often unnoticed (De Andrade et al. 2001; Uchida et al. 2009). While the interforaminal area is considered to be a safe region to perform periodontal surgery with minimal morbidity, the area should be carefully assessed, especially if there is a large incisive branch. In cases where a patient may have a large incisive branch in close proximity to the surgical area, he or she may experience discomfort during the surgery or may experience postoperative pain. Consequently, careful consideration at the surgical treatment planning stage should be performed to assess the precise size and location of the incisive branch before performing periodontal surgical procedures deeply in the interforaminal area.

2.3.4 The Lingual Nerve

The lingual nerve is a branch of the mandibular nerve, which enters the oral cavity above the posterior edge of the mylohyoid muscle close to the third molar tooth proceeding on the surface of the hypoglossal muscle to enter the floor of the mouth and the tongue.

2.3.4.1 Variations in the Anatomic Configuration of the Lingual Nerve

When reflecting the lingual gingiva in the posterior mandibular region, the periodontist needs to carefully elevate the flap to avoid injuring the lingual nerve. The lingual nerve provides sensation to the mucous membranes of the anterior two-thirds of the tongue and the lingual tissues. According to Behnia et al. (2000), the lingual nerve is located 3 mm apical to the osseous crest and 2 mm horizontally from the lingual cortical plate in the flap . However, this position and location of the lingual nerve can vary among individuals. In 15–20% of cases, the nerve may be situated at or above the crest of bone, lingual to the mandibular third molars (Pogrel and Goldman 2004). Alternatively, and in 22% of the cases, the lingual nerve may contact the lingual cortical plate (Behnia et al. 2000). When performing surgery in this area, it is important for the clinician to ascertain the precise location of the lingual nerve, protect it in the flap with a blunt-ended elevator, minimize trauma when reflecting the flap to avoid mechanical injury to the tissue, and avoid making lingual vertical incisions (Greenstein et al. 2008).

2.3.5 Buccal Nerve

The (long) buccal nerve is a branch of the anterior division of the mandibular nerve. In general, the buccal nerve courses in between the two heads of the lateral pterygoid muscle, anterior to the tendon of the temporalis muscle, to connect with the buccal branch of the facial nerve on the surface of the buccinator muscle.

2.3.5.1 Variations in the Anatomic Configuration of the Buccal Nerve

A variation characterized as an aberrant nerve arising from a small foramen in the retromolar fossa is known as Turner's variation of the buccal nerve (Turner 1864). If this variation exists, surgical trauma can cause injury to the nerve and cause paresthesia to the adjacent soft tissue.

References

Anderson LC, Kosinski TF, Mentag PJ (1991) A review of the intraosseous course of the nerves of the mandible. J Oral Implantol 17:394–403

Behnia H, Kheradvar A, Shahrokhi M (2000) An anatomic study of the lingual nerve in the third molar region. J Oral Maxillofac Surg 58:649–651

Boyne PJ, James RA (1980) Grafting of the maxillary sinus floor with autogenous marrow and bone. J Oral Surg 38:613–616

Dario LJ (2002) Implant placement above a bifurcated mandibular canal: a case report. Implant Dent 11:258–261

De Andrade E, Otomo-Corgel J, Pucher J et al (2001) The intraosseous course of the mandibular incisive nerve in the mandibular symphysis. Int J Periodontics Restorative Dent 21:591–597

Fishel D, Buchner A, Hershkowith A et al (1976) Roentgenologic study of the mental foramen. Oral Surg Oral Med Oral Pathol 41:682–686

Fu JH, Hasso DG, Yeh CY et al (2011) The accuracy of identifying the greater palatine neurovascular bundle: a cadaver study. J Periodontol 82:1000–1006

Greenstein G, Cavallaro J, Tarnow D (2008) Practical application of anatomy for the dental implant surgeon. J Periodontol 79:1833–1846

Harris RJ, Miller R, Miller LH et al (2005) Complications with surgical procedures utilizing connective tissue grafts: a follow-up of 500 consecutively treated cases. Int J Periodontics Restorative Dent 25:449–459

Kim MJ, Jung UW, Kim CS et al (2006) Maxillary sinus septa: prevalence, height, location, and morphology. A reformatted computed tomography scan analysis. J Periodontol 77:903–908

Klosek SK, Rungruang T (2009) Anatomical study of the greater palatine artery and related structures of the palatal vault: considerations for palate as subepithelial connective tissue graft donor site. Surg Radiol Anat 31:245–250

Krennmair G, Ulm CW, Lugmayr H et al (1999) The incidence, location, and height of maxillary sinus septa in the edentulous and dentate maxilla. J Oral Maxillofac Surg 57:667–671
Mogensen C, Tos M (1977) Quantitative histology of the maxillary sinus. Rhinology 15:129–140
Ngeow WC, Yuzawati Y (2003) The location of the mental foramen in a selected Malay population. J Oral Sci 45:171–175
Pogrel MA, Goldman KE (2004) Lingual flap retraction for third molar removal. J Oral Maxillofac Surg 62:1125–1130
Reiser GM, Bruno JF, Mahan PE et al (1996) The subepithelial connective tissue graft palatal donor site: anatomic considerations for surgeons. Int J Periodontics Restorative Dent 16:130–137
Tidwell JK, Blijdorp PA, Stoelinga PJ et al (1992) Composite grafting of the maxillary sinus for placement of endosteal implants. A preliminary report of 48 patients. Int J Oral Maxillofac Surg 21:204–209
Tomaszewska IM, Tomaszewski KA, Kmiotek EK et al (2014) Anatomical landmarks for the localization of the greater palatine foramen – a study of 1200 head CTs, 150 dry skulls, systematic review of literature and meta-analysis. J Anat 225:419–435
Turner W (1864) On some variations in the arrangement of the nerves of the human body. Nat Hist Rev 4:612–617
Uchida Y, Noguchi N, Goto M et al (2009) Measurement of anterior loop length for the mandibular canal and diameter of the mandibular incisive canal to avoid nerve damage when installing endosseous implants in the interforaminal region: second attempt introducing cone beam computed tomography. J Oral Maxillofac Surg 67:744–750
Underwood AS (1910) An inquiry into the anatomy and pathology of the maxillary sinus. J Anat Physiol 44:354–369
Wang TM, Shih C, Liu JC et al (1986) A clinical and anatomical study of the location of the mental foramen in adult Chinese mandibles. Acta Anat (Basel) 126:29–33

Anatomical Issues Related to Endodontics

3

Charles S. Solomon and Sahng G. Kim

Anatomical issues play a very crucial role in the everyday practice of endodontics. Rather than be a subject to be endured and soon forgotten in dental school, an understanding of anatomy is critical to the endodontist, both in reaching the proper diagnosis and then in executing the proper techniques to reach the most successful outcomes.

There are two main areas of anatomy that should concern the endodontist. One is the internal anatomy of the root canal space. The other is the anatomy of the maxilla and mandible where the teeth reside.

3.1 Anatomy of the Root Canal Space as it Affects Endodontic Outcomes

For educational and demonstrative reasons, the root canal space is always portrayed two-dimensionally, as a smooth, straight conical-shaped canal that tapers to a point at its apical end. In reality, this idealized version is never really the case. In actuality like fingerprints, no two spaces are ever the same. Any given tooth can have extra canals, several foramina, and multiple fins, isthmuses, and deltas (Vertucci 1984). Accessory and lateral canals are not uncommon, particularly in the furcal and apical regions (De Deus 1975; Burch and Hulen 1974). Most root canals are curved as they travel apically. The cone beam computed tomography (CBCT) rapidly becoming an essential part of the endodontist's armamentarium helps show, three-dimensionally, the complexities of the root canal space (Ratanajirasut et al. 2018). The understanding of this space is much more than a mere academic exercise for an endodontist. It

C. S. Solomon (✉) · S. G. Kim
Division of Endodontics, Columbia University College of Dental Medicine, New York, NY, USA
e-mail: css5@cumc.columbia.edu; sgk2114@cumc.columbia.edu

J. Iwanaga, R. S. Tubbs (eds.), *Anatomical Variations in Clinical Dentistry*,
https://doi.org/10.1007/978-3-319-97961-8_3

affects his technical and biological treatment of the root canal and ultimately his outcomes, both successes and failures.

The orifice of the canal must be enlarged by the endodontist to allow access for the instruments, irrigants, and medicaments to effectively clean and shape the root canal space (Lim and Webber 1985). A large array of different cleaning systems is available, using both hand and rotary files. These instruments all are machined to an idealized tapered shape and cannot duplicate the shape of the root canal space. Micro-computed tomography studies show that significant areas of the canal walls, as much as 40%, are untouched by the root canal file (Gekelman et al. 2009). Recently irregularly shaped files have been developed to produce a whipping action, in an attempt to scrub more of the canal walls (XP-3D file system, Self-Adjusting file system) (Lacerda et al. 2017). Untouched walls obviously present a problem, particularly in necrotic cases where bacteria are present and often tenaciously attached to the walls in biofilms (Ricucci et al. 2009; Vieira et al. 2012; Chávez de Paz et al. 2007). Therefore, instrumentation alone cannot produce a sterile canal. We must rely heavily on irrigants such as sodium hypochlorite to dissolve organic tissue and kill microorganisms, as well as on chelating agents such as EDTA to decalcify inorganic tissue (Mohammadi 2008; Hulsmann et al. 2003; Yang et al. 2016). Between visits, we use intracanal medicaments, such as calcium hydroxide to further render the space free of microorganisms (Sjogren et al. 1991; Katebzadeh et al. 2000).

Finally, the endodontist is acutely aware of the need to preserve the narrow constriction of the root canal space, the apical foramen, which usually lies about 0.5 mm from the radiographic apex (Kuttler 1955; Ricucci 1998). Because this terminus of the root canal space is the narrowest part of the canal, it allows the endodontist to obturate the space with a well-condensed filling that can flow into all the fins, isthmuses, and deltas, without extruding apically (Ponce and Vilar Fernandez 2003; Jarrett et al. 2004). When this apical constriction is compromised iatrogenically by overzealous instrumentation or by periradicular infection that can resorb the end of the root, the opportunity to place a well-condensed root filling is greatly reduced. In essence, the natural funnel shape of a root canal space has been replaced by a tubular shape, and the chances of pushing the filling material through the apical foramen are much greater.

3.1.1 Local Anesthesia

Attaining profound anesthesia is an essential element in successful endodontic therapy. The endodontist must have a detailed understanding of the anatomy of the oral cavity in order to produce an effective level of anesthesia. Knowing the location of the pathways of the nerves and blood vessels as they course through the maxilla and mandible is critical in producing adequate anesthesia without causing paresthesias that can lead to permanent injury. While teeth in the maxilla can be numbed with a simple infiltration, where the needle just penetrates the mucosa a few millimeters to the periosteum, the teeth in the mandible are situated in denser bone with a thick cortical plate and cannot be numbed with an infiltration technique. To anesthetize mandibular teeth, we must block the nerve bundles as they emerge from their bony channels. In particular, we must know the pathway of the inferior alveolar nerve and the locations of the mental foramen and mandibular foramen. To anesthetize larger maxillary areas when apical surgery is planned, infiltration may not be adequate. We must be familiar with the infraorbital nerve, greater palatine nerve, nasopalatine nerve, and the branches of the superior alveolar nerves and know the locations of the foramina from which they exit their bony channels in order to be able to block these nerve bundles. To effect block anesthesia, the needle may have to pass through tissue for an inch or more, such as with the inferior alveolar nerve in the mandible. In such cases we must attempt to bypass the muscles of mastication, such as the masseter and pterygoid. When a needle passes through the muscle instead of bypassing it, postoperative soreness and spasm leading to trismus are possible. Understanding the insertions and directions of the muscles in the oral cavity is therefore essential. In rare instances the anesthetic needle has been known to penetrate the foramen, causing mechanical injury to the nerve bundle and creating disabling sensory disturbances such as pain, hypoesthesia, paresthesia, and dysesthesia of the lower lip and chin (Donoff and Colin 1990; Vallerand 1992; Pogrel and Thamby 2000; Ogle and Mahjoubi 2012). The mechanism of this untoward event is believed to be the needle piercing the nerve bundle and creating a capillary bleed in the confined space of the bony canal, which then increases the pressure, causing destruction of nerve tissue (Donoff and Colin 1990; Vallerand 1992; Pogrel and Thamby 2000; Kitagawa et al. 2004). The lingual nerve is injured twice as often as the inferior alveolar nerve because it is unifascicular as opposed to the inferior alveolar nerve which is multifascicular in the area of the lingula (Pogrel et al. 2003; Moore and Haas 2010). If a microsurgeon is consulted in a timely fashion, there is a reasonable chance that he can decompress the area by decorticating the inferior alveolar nerve, removing the withered nerve tissue, and reattaching the nerve bundle, thus restoring sensation to the patient. Time is critical in this situation. Pogrel (2007; 2002) has shown that he achieved 100% success when he performed the surgery within 48 h of injury, but his success diminished significantly as time between injury and surgery increased.

3.1.2 Nonsurgical Endodontic Treatment

While the incidence of permanent paresthesia from an anesthetic block is exceedingly rare, it is far more common from iatrogenic injury following nonsurgical endodontic treatment. The endodontist may inadvertently push the endodontic file beyond the apex, and/or force the endodontic irrigant and paste beyond the apex, into the inferior alveolar nerve bundle, thus creating paresthesia, due either to chemical neurotoxicity or mechanical pressure. This compression of the artery in the nerve bundle results in increased vascular permeability and edema and decreases oxygen to the nerve bundle (Pogrel 2007).

These nerve injuries occur when the apex of the tooth comes into direct contact with the nerve bundle. While a periapical radiograph, being two dimensional, can never indicate the true proximity of the apex to the mandibular canal, the CBCT can accurately determine this relationship. This knowledge could help the endodontist avoid the paresthesia pitfalls. When the CBCT shows that the tooth apex and the mandibular canal are contiguous, he could modify his techniques by instrumenting and then filling the canal short of the apex, using more viscous pastes and smaller amounts. The critical areas for this paresthesia occurrence, where the apices are in close proximity to the mandibular canal, are the mandibular second and third molars and the mandibular bicuspids which are often in contact with the inferior alveolar nerve as it emerges from the mental foramen (Moiseiwitsch 1995) (Figs. 3.1 and 3.2).

In nonsurgical endodontic treatment of the maxillary teeth, the nerve bundles are of less a factor because they are a considerable distance from the apices of the teeth. The major endodontic concern in the maxilla is the maxillary sinus. The sinus dips down between the roots of the posterior teeth, often with less than 1 mm separation

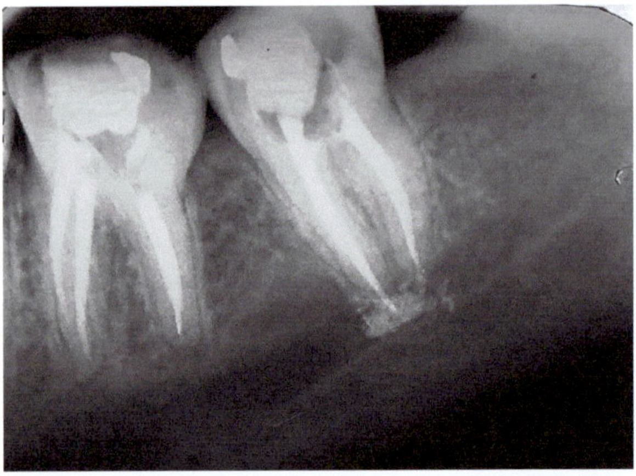

Fig. 3.1 Apices of roots in direct contact with inferior alveolar nerve. Apical constriction of the mesial root of left mandibular second molar was compromised iatrogenically, allowing sealer to flow into inferior alveolar nerve, causing permanent paresthesia

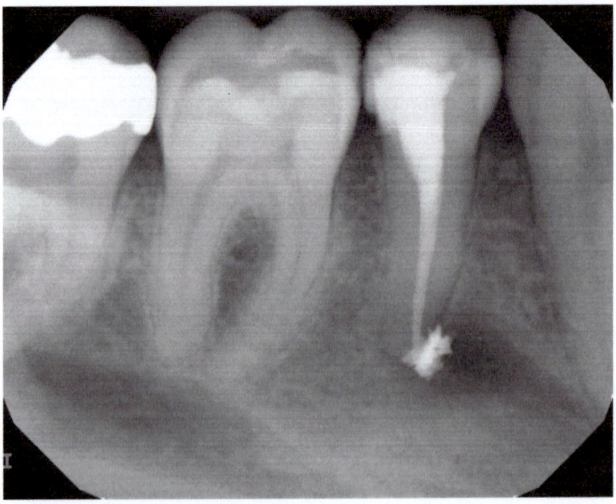

Fig. 3.2 Apex of the root of right mandibular second premolar in direct contact with mental foramen. Apical constriction of mandibular premolar was compromised iatrogenically, allowing sealer to flow into inferior alveolar nerve as it exits mental foramen, causing permanent paresthesia

(Hauman et al. 2002). The Schneiderian membrane, the anatomical barrier between the root and sinus, is often compromised by apical inflammation and infection. Instrument perforation of the sinus is not uncommon (Freedman and Horowitz 1999). Sound endodontic therapy under sterile conditions, however, has been shown to lead to excellent long-term outcomes, despite instrument perforations of the sinus. It is not uncommon for a chronically infected posterior maxillary tooth to drain into the sinus and cause mucositis (sinus congestion), evident in a CBCT. This can be nonsurgically reversed with effective endodontic therapy (Fig. 3.3).

3.1.3 Surgical Endodontic Treatment

The same anatomical landmarks just reviewed have even greater immediacy in surgery because the scalpel and curette have the potential for much more irreversible damage. In planning and then executing the flap for mandibular surgery, the endodontist must be aware of the mental foramen. By identifying the neurovascular bundle as it emerges from the mental foramen, he is then able to avoid it. His anterior-releasing incision, therefore, should be anterior to the foramen. As the surgeon goes further posterior, the roots tend to curve more lingually (second and third molars), thus complicating his access to the apices. Compound this with a shallow vestibule and thick horizontal cortical plate of bone, as is often the case, and the prognosis for a good outcome diminishes dramatically. A CBCT is therefore critical in treatment planning for surgery in the posterior mandible.

The maxillary sinus is the main obstacle to successful endodontic surgery in the maxilla. Fortunately, the issues that prevent nonsurgical treatment of maxillary molars are predominantly related to the buccal roots. The canals in the buccal roots are curved, narrow, and easily obstructed. The palatal canal is the largest, straightest, and most amenable to nonsurgical retreatment. In performing apical surgery on the buccal roots, perforation of the sinus is not uncommon, occurring as often as 50% of the time (Rud and Rud 1998). These perforations will heal and close with new bone and membrane regeneration. Studies show that sinus perforations during surgery do not have a deleterious effect on the outcome (Watzek et al. 1997). The

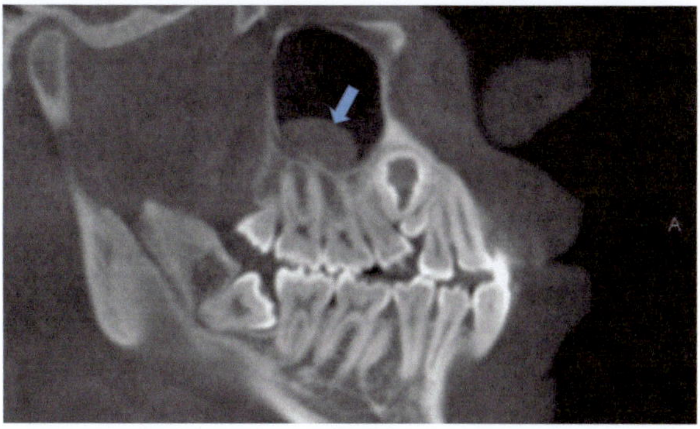

Fig. 3.3 Arrow pointing to the level of mucosal thickening on the floor of sinus, related to roots with necrotic (infected) root canals that extend to the sinus

major concern for the surgeon is to avoid dropping foreign bodies related to the retrograde apical seal or infected root fragments into the sinus. To avoid this situation, he must temporarily pack the perforation site while sealing the apex. The vertical releasing incisions in all surgeries must be placed away from muscle attachments and mesially and/or distally far enough away from the lesion to ensure that suturing will be done on sound bone. A vertical incision is preferred over an angled incision because it cuts fewer blood vessels, thus allowing an adequate blood flow for the subsequent wound healing.

When the palatal root canal is infected and cannot be negotiated, a surgical retrograde approach is the only option. The buccal approach is fraught with problems because the sinus floor might be opened greater than 1 cm in reaching the palatal canal. A palatal approach might be more prudent if a CBCT indicates that the apex of the root is close to the palate. The releasing flap incision on the palate must be made anteriorly to avoid the greater palatine artery that emerges from the greater palatine foramen. Research at Columbia is currently underway where, using a CBCT, a template is constructed that would guide a 5 mm trephan bur through the palate to punch a hole to the apex, without a palatal flap.

There are less potential complications in anterior maxillary surgery; however, the anterior nasal spine and the floor of the nasal cavity must be identified and avoided, particularly in dealing with teeth with long roots. Apical surgery on mandibular incisors can be challenging, especially with shallow vestibules. These roots tend to curve lingually, making an apical preparation perpendicular to the long axis of the tooth, exceedingly difficult to achieve. Mandibular incisors are close and often crowded, and it is challenging to perform an apicoectomy without injuring the adjacent teeth.

3.1.4 Fascial Spaces, Compartments, or Clefts

Finally, the practicing endodontist must be aware of the fascial spaces in the head and neck because these are the potential pathways by which infection may spread from the tooth apex to vital organs with significant morbidity and mortality. Ludwig's angina, cervicofacial necrotizing fasciitis, descending necrotizing mediastinitis, cavernous sinus thrombosis, and brain abscess are orofacial infections that can have their origin in a periapical abscess (Lazow 2000; Izzo and Lazow 2001). Deep fascia is a connective tissue around different structures such as bone and muscle. Fascial compartments are not empty but rather are potential spaces for infection to pass through the fascial layers. The clinically important fascial spaces in the face are the sublingual, submandibular, and submental in the floor of the mouth, the masticator and pterygomandibular spaces near the ramus, and the buccal space (Flynn 1994). An abscess may spread between the layers of cervical fascia to the retropharyngeal space and then ascend to the skull or descend into the mediastinum.

The endodontist must appreciate the significance of these potential fascial pathways that can spread an infection systemically. Not knowing the virulence of the microorganism or the host's resistance to infection, in any specific situation, the endodontist should stay in close contact with the patient for the first 24–48 h after initiating treatment of periapical infection.

We have attempted to outline the many aspects of anatomy that come into play for the endodontist in his daily treatment of patients.

References

Burch JG, Hulen S (1974) A study of the presence of accessory foramina and the topography of molar furcations. Oral Surg Oral Med Oral Path 38:451–455

Chávez de Paz LE, Bergenholtz G, Dahlén G et al (2007) Response to alkaline stress by root canal bacteria in biofilms. Int Endod J 40:344–355

De Deus QD (1975) Frequency, location, and direction of the lateral, secondary, and accessory canals. J Endod 1:361–366

Donoff RB, Colin W (1990) Neurologic complications of oral and maxillofacial surgery. Oral Maxillofac Surg Clin 2:453–462

Flynn TR (1994) Anatomy and surgery of deep fascial space infections of the head and neck. In: JPW K (ed) OMS knowledge update, vol 1. Rosemont IL, AAOMS, p 104

Freedman A, Horowitz I (1999) Complications after apicoectomy in maxillary premolar and molar teeth. Int J Oral Maxillofac Surg 28:192

Gekelman D, Ramamurthy R, Mirfarsi S et al (2009) Rotary nickel-titanium GT and ProTaper files for root canal shaping by novice operators: a radiographic and micro-computed tomography evaluation. J Endod 35:1584–1588

Hauman CH, Chandler NP, Tong DC (2002) Endodontic implications of the maxillary sinus: a review. Int Endod J 35:127

Hulsmann M, Heckendorff M, Lennon A (2003) Chelating agents in root canal treatment: mode of action and indications for their use. Int Endod J 36:810

Izzo SR, Lazow SK (2001) Early diagnosis and surgical management of necrotizing fasciitis, mediastinitis, and brain abscess. In: Piecuch JF (ed) OMS knowledge update, vol 3. AAOMS, Rosemont, IL, p 102

Jarrett IS, Marx D, Covey D et al (2004) Percentage of canals filled in apical cross sections—an in vitro study of seven obturation techniques. Int Endod J 37:392–398

Katebzadeh N, Sigurdsson A, Trope M (2000) Radiographic evaluation of periapical healing after obturation of infected root canals: an in vivo study. Int Endod J 33:60–66

Kitagawa N, Oda M, Totoki T (2004) Possible mechanism of irreversible nerve injury caused by local anesthetics: detergent properties of local anesthetics and membrane disruption. Anesthesiology 100:962–967

Kuttler Y (1955) Microscopic investigation of root apexes. J Am Dent Assoc 50:544

Lacerda MFLS, Marceliano-Alves MF, Pérez AR et al (2017) Cleaning and shaping oval canals with 3 instrumentation systems: a correlative micro-computed tomographic and histologic study. J Endod 43(11):1878–1884

Lazow SK (2000) Necrotizing fasciitis and mediastinitis. Atlas Oral Maxillofac Surg Clin North Am 8:101

Lim KC, Webber J (1985) The effect of root canal preparation on the shape of the curved root canal. Int Endod J 18:233–239

Mohammadi Z (2008) Sodium hypochlorite in endodontics: an update review. Int Dent J 58:329

Moiseiwitsch JR (1995) Avoiding the mental foramen during periapical surgery. J Endod 21:340

Moore PA, Haas DA (2010) Paresthesias in dentistry. Dent Clin N Am 54:715–730

Ogle OE, Mahjoubi G (2012) Local anesthesia: agents, techniques, and complications. Dent Clin N Am 56:133–148

Pogrel MA (2002) The results of microneurosurgery of the inferior alveolar and lingual nerve. J Oral Maxillofac Surg 60:485–489

Pogrel MA (2007) Damage to the inferior alveolar nerve as the result of root canal therapy. JADA 138:65–69
Pogrel MA, Thamby S (2000) Permanent nerve involvement resulting from inferior alveolar nerve blocks. JADA 131:901–906
Pogrel MA, Schmidt BL, Sambajon V (2003) Lingual nerve damage due to inferior alveolar nerve blocks. JADA 134:195–198
Ponce EH, Vilar Fernandez JA (2003) The cemento-dentino-canal junction, the apical foramen, and the apical constriction: evaluation by optical microscopy. J Endod 29:214
Ratanajirasut R, Panichuttra A, Panmekiate S (2018) A cone-beam computed tomographic study of root and canal morphology of maxillary first and second permanent molars in a Thai population. J Endod 44:56–61
Ricucci D (1998) Apical limit of root canal instrumentation and obturation, part 1. Literature review. Int Endod J 31:384–393
Ricucci D, Siqueira JF Jr, Bate AL et al (2009) Histologic investigation of root canal-treated teeth with apical periodontitis: a retrospective study from twenty-four patients. J Endod 35:493–502
Rud J, Rud V (1998) Surgical endodontics of upper molars: relation to the maxillary sinus and operation in acute state of infection. J Endod 24:260
Sjogren U, Figdor D, Spangberg L, Sundqvist G (1991) The antimicrobial effect of calcium hydroxide as a short- term intracanal dressing. Int Endod J 24:119
Vallerand WP (1992) Peripheral trigeminal nerve injuries. N Y State Dent J 8:27–30
Vertucci FJ (1984) Root canal anatomy of the human permanent teeth. Oral Surg Oral Med Oral Pathol 58:589–599
Vieira AR, Siqueira JF Jr, Ricucci D et al (2012) Dentinal tubule infection as the cause of recurrent disease and late endodontic treatment failure: a case report. J Endod 38:250–254
Watzek G, Bernhart T, Ulm C (1997) Complications of sinus perforations and their management in endodontics. Dent Clin N Am 41:563
Yang Y, Shen Y, Wang Z et al (2016) Evaluation of the susceptibility of multispecies biofilms in dentinal tubules to disinfecting solutions. J Endod 42(8):1246–1250

Part II

Mandible

Anatomy and Variations of the Pterygomandibular Space

4

Iwona M. Tomaszewska, Matthew J. Graves, Marcin Lipski, and Jerzy A. Walocha

4.1 Introduction

The pterygomandibular space (PM) is a space largely composed of connective tissue and muscle but also contains several neurovascular structures (Shields 1977). It is triangular in shape and is bounded by the medial surface of the mandibular ramus laterally (Huelke 1973; Lipski et al. 2013a; Shaw and Fierst 1988). The medial pterygoid muscle and associated fascia confine the space medially and the lateral pterygoid superiorly. The parotid gland lies posteriorly, and anteriorly, the buccinator, superior pharyngeal constrictors, and pterygomandibular raphe can be found (Lipski et al. 2013a; Shaw and Fierst 1988). Of note, the inferior alveolar nerve, artery, and vein, along with the lingual nerve, traverse the PM. Detailed knowledge of the PM is important for the successful inferior alveolar nerve blockade (Fig. 4.1) (Khoury et al. 2010). Awareness of the varying relationships of neurovascular structures to both osseous and connective tissue landmarks in the PM is essential for avoiding iatrogenic injury during nerve blocks and dental procedures in this specific region. In this chapter, the most common anatomy of the PM will be reviewed, along with common variants and their clinical implications.

4.2 Boundaries, Musculature, and Fascia

The PM contains several neurovascular structures; however there are also an abundance of soft tissue structures (Fig. 4.2). The medial border of the PM is composed of the interpterygoid fascia laterally abutting the medial pterygoid muscle (Barker

I. M. Tomaszewska (✉) · M. J. Graves · M. Lipski · J. A. Walocha
Department of Anatomy, Jagiellonian University Medical College, Krakow, Poland

J. Iwanaga, R. S. Tubbs (eds.), *Anatomical Variations in Clinical Dentistry*,
https://doi.org/10.1007/978-3-319-97961-8_4

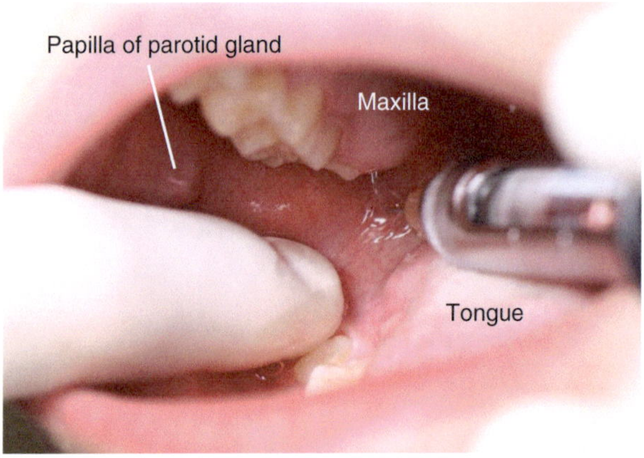

Fig. 4.1 Inferior alveolar nerve block

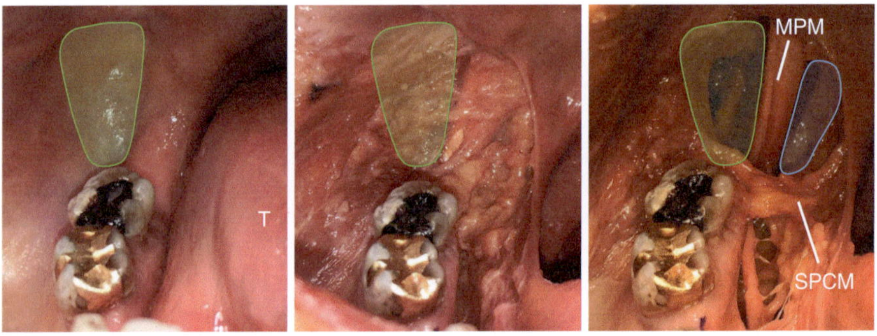

Fig. 4.2 Intraoral dissection of pterygomandibular space (light green) and parapharyngeal space (light blue). *MPM* medial pterygoid muscle, *SPCM* superior pharyngeal constrictor muscle, *T* tongue

and Davies 1972). The medial pterygoid muscle is one of the masticator muscles. The interpterygoid fascia continues superiorly to cover the medial border of the lateral pterygoid muscle, the superior border of the PM (Barker and Davies 1972) (Fig. 4.3). The lateral pterygoid is the only muscle of mastication associated with opening the mouth. The lateral wall of the PM is made up of the medial surface of the mandibular ramus, and notable structures in this review are the lingula, sulcus colli, and temporal crest (Barker and Davies 1972). The posterior boundary is composed of the stylomandibular membrane (located between the syloid process and posterior border of the mandible) and subsequent medial portion of the parotid gland (Barker and Davies 1972). The PM is bounded anteriorly by several different structures. Most proximal to the inferior alveolar neurovascular bundle are the temporalis muscle tendons followed by the temporopterygoid fascia (Barker and Davies 1972). The ligamentous pterygomandibular raphe serves as an attachment for the buccinator muscle anteriorly and the superior pharyngeal constrictor posteriorly (Barker and Davies 1972).

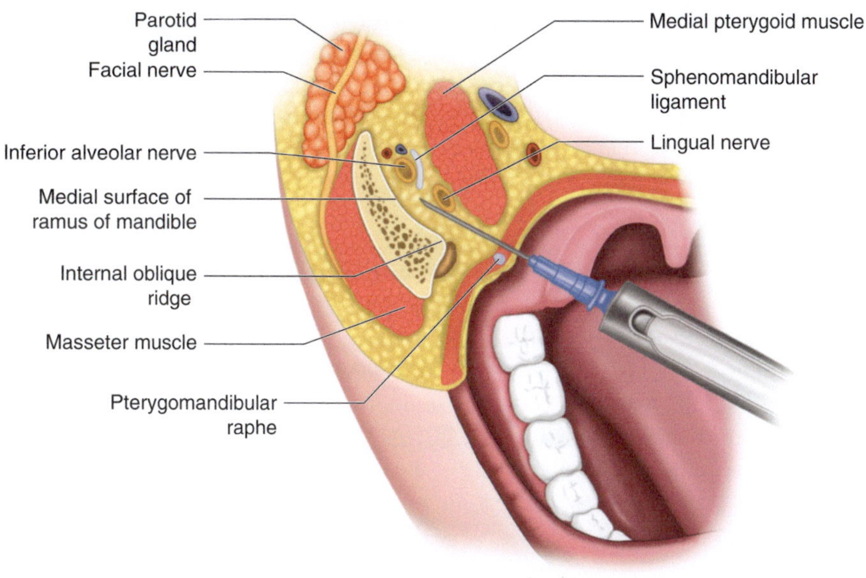

Fig. 4.3 Schematic drawing of pterygomandibular space (superior view)

Fig. 4.4 Lateral course right maxillary artery. *IAA* inferior alveolar artery, *MA* maxillary artery, *LPM* lateral pterygoid muscle, *P* posterior, *A* anterior

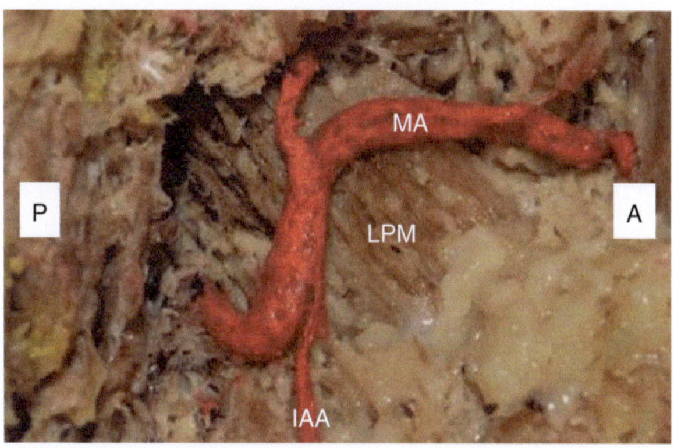

4.3 Maxillary Artery

The maxillary artery (MA) in some cases has been visualized within the PM. When present in the PM, it can be found within the interpterygoid fascia covering the lateral pterygoid muscle (Barker and Davies 1972). From a surgical point of view, the relationship between the lateral pterygoid muscle and the MA is crucial, especially regarding the bilateral sagittal split osteotomy. When the MA courses lateral to the lateral pterygoid muscle, the MA is in high risk of injury when the medial horizontal bony incision of the ramus is made (Fig. 4.4). Incidence of the internal course of the MA has been reported as 3.6–10.1% in Japanese, 44.7% in the United Kingdom, 45.6% of Caucasians in the United States, and 91.5% in Australians (Iwanaga et al. 2017a).

Cases of anatomic variants involving the MA have been reported in the literature. One instance of such a variant was reported in a cadaver with a divided mandibular nerve entrapping the MA via two branches that reunited into the inferior alveolar nerve distally (Anil et al. 2003). The entrapment of the neural tissue by the MA may potentially lead to pain and numbness in the sensory distribution of that nerve due to the arterial pulsations or direct compression (Wolf et al. 2016).

4.4 Inferior Alveolar Nerve, Artery, and Vein

The inferior alveolar neurovascular bundle is always closely related to the lingula of the mandible, localizing most commonly in a posterolateral position (Khoury et al. 2010; Lipski et al. 2013b). It contains the inferior alveolar nerve, artery, and vein within a fibrous sheath (Khoury et al. 2010). The inferior alveolar neurovascular bundle courses along the medial aspect of the mandibular ramus and is, on average, 60% of the distance from the anterior border of the ramus compared to total ramus width (Khoury et al. 2010).

4.4.1 Inferior Alveolar Nerve

The inferior alveolar nerve (IAN) is a sensory nerve branch of the mandibular nerve (CN V_3) providing the somatosensory innervation of the mandibular teeth. The lingual nerve and the IAN divide deep to the lateral pterygoid muscle and enter the PM as separate structures from the lateral surface of the medial pterygoid (Barker and Davies 1972). The IAN is noted macroscopically as a single nerve trunk; however it is actually comprised of two individual nerves: the mental nerve and the incisive branches. Grossly, the IAN's external morphology is variable with approximately one-third of nerves fitting into each category of round, oval, and flat shaped, 2 mm above the lingula (Tan et al. 2014). The IAN at the level of the lingula is found in the anterior part of the sulcus colli as it courses toward its ultimate entrance into the mandibular foramen (Khoury et al. 2010) (Fig. 4.5). The lingula, more specifically the slightly superior to the temporal crest, is the most anterior bony ridge of the sulcus colli and may provide a level of direct protection to the IAN during an IAN blockade (Barker and Davies 1972; Jorgensen and Hayden 1967). Numerous techniques have been developed to anesthetize the mandible in recent years; however the direct approach of the IAN block remains a staple for many clinicians (Malamed 2004). Failed block is the most common problem with an IAN blockade and can approach 20% in certain cases (Kaufman et al. 1984). Depending on the clinician's techniques and a patient's anatomical variants, the lingual nerve may be concomitantly anesthetized by the IAN blockade (Khoury et al. 2010). This dual anesthetization may be desirable in certain cases, and clinicians should be aware that the distance between the IAN and the LN is, on average, 5.3–8.5 mm and may aid in the approximation of needle withdrawal required (Khoury et al. 2010; Iwanaga et al. 2018). A notable variant is the presence of communicating branches between IANs

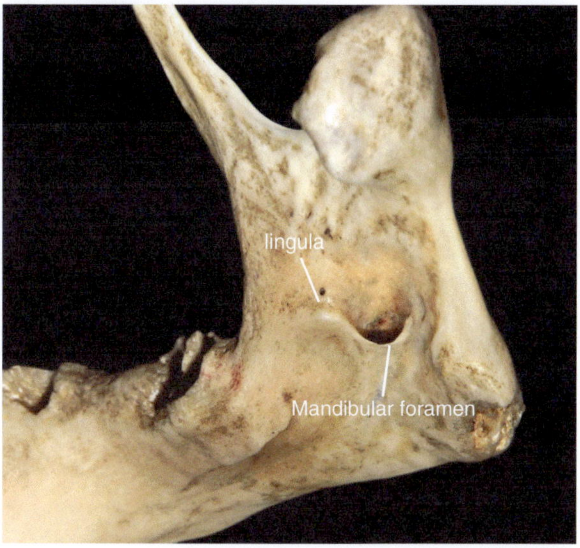

Fig. 4.5 Lingula and mandibular foramen

and LNs (Wolf et al. 2016). In addition to physical barriers, anatomic variations of the region may also construe the classically portrayed anatomy. Structures, such as the accessory mandibular foramen, have given rise to bifid IANs (Arias et al. 1967; Schejtman et al. 1967). Surgical procedures themselves may inflict harm to the IAN as well, particularly those involving the third mandibular molar (Matani et al. 2014). The incidence of IAN damage during third molar tooth surgery has been reported in up to 8% (Matani et al. 2014; Nakagawa et al. 2007; Blaeser et al. 2003). These factors may contribute to the high failure rate of the IAN blockade (Malamed 2004). Collateral sensory innervation of the mandibular teeth from nerves such as the nerve to the mylohyoid may potentially play a role in the unpredictable nature of IAN blockades (Frommer et al. 1972). Detailed knowledge of the arrangement and depth of structures within the PM may permit more effective local anesthetic administration.

4.4.2 Inferior Alveolar Artery

The inferior alveolar artery (IAA) originates from the maxillary artery and courses with the IAN (Shaw and Fierst 1988) (Fig. 4.6). The IAA is most commonly (79% of cases) located posteriorly or posterolaterally to the IAN (Khoury et al. 2010) and is the most common spatial arrangement historically (Barker and Davies 1972; Sicher and DuBrul 1975; Murphy and Grundy 1969). The positioning of the IAA within the PM is procedurally advantageous as it may result in fewer vascular aspirations and unnecessary vascular trauma during an IAN blockade (Khoury et al. 2010). Case reports have demonstrated that hematoma may occur as a complication from an IAN blockade (Traeger 1979). Hematoma formation from a small vessel

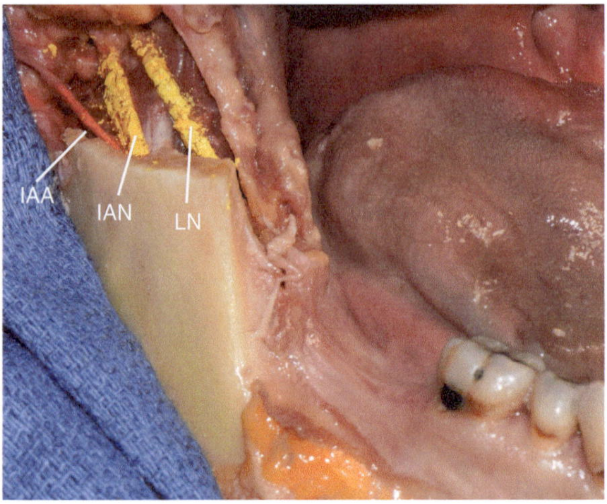

Fig. 4.6 Right inferior alveolar nerve (IAN) and inferior alveolar artery (IAA). Upper part of mandibular ramus removed. *LN* lingual nerve

such as the IAA is possible, however more likely from large structures such as the maxillary veins and pterygoid venous plexus, with high needle placement (Shaw and Fierst 1988; Traeger 1979). Aberrant origins of the IAA have been reported in the literature such as a case of external carotid artery origin (Jergenson et al. 2005). The fact that the inferior alveolar vasculature rarely lies medial to the IAN may reduce the incidence of intravascular injection of local anesthetics (Khoury et al. 2010). It should be noted however that approximately 15% of IAN blocks are aspiration positive (Taghavi Zenouz et al. 2008).

4.4.3 Inferior Alveolar Vein

The inferior alveolar vein (IAV) is most commonly found as the most posterior structure of the inferior alveolar neurovascular bundle (Khoury et al. 2010). That is, in 71% of cases, it is found posteriorly or posterolaterally to the IAA (Khoury et al. 2010). The IAV eventually empties into the pterygoid venous plexus, which is closely associated with the lateral pterygoid muscle (Khoury et al. 2011). There is also the possibility that the pterygoid venous plexus may extend inferiorly in place of the IAV serving as a plexus surrounding the IAA (Murphy and Grundy 1969). The IAV was historically portrayed as a single structure. However, in recent cadaveric examinations, the presence of two veins was the most prevalent pattern (Khoury et al. 2010). It should be noted that due to the IAV often being the most posterior structure within the inferior alveolar neurovascular sheath, it may be afforded fewer of the osseous protections than its arterial and neural counterparts (Khoury et al. 2010). Procedurally, it is recommended withdrawing anesthetic needles 1–2 mm prior to performing an aspiration as to avoid pinning of the IAV to the mandible (Khoury et al. 2010).

4.5 Lingual Nerve

The lingual nerve (LN) is a branch of the mandibular nerve (Tan et al. 2014), but it also receives somatic sensory innervation from the anterior two-thirds of the tongue along with a small subset of special sensory taste fibers and parasympathetics originating from the facial nerve (CN VII) (Takezawa and Kageyama 2015). Additionally, it is believed that the lingual nerve carries proprioceptive information regarding the position of the tongue (Barker and Davies 1972). The LN, when traced more proximally from the PM, can be viewed joining together with the IAN at their common origin of the mandibular nerve (Fig. 4.7) (Shaw and Fierst 1988). From the posterior aspect, the chorda tympani contributes special sensory fibers from the facial nerve (Shaw and Fierst 1988). The LN courses through the PM along the anterolateral aspect of the medial pterygoid muscle continuing inferiorly to enter the submandibular area (Al-Amery et al. 2016). The external morphology of the LN is similar to that of the IAN with approximately one-third of nerves fitting into each category of oval, flat, and circular shaped at a level of 2 mm above the lingula (Tan et al.

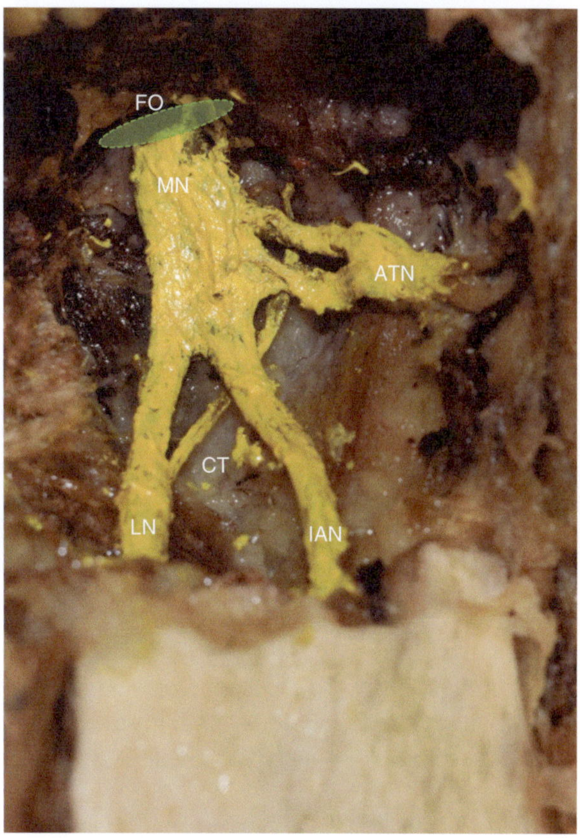

Fig. 4.7 Origin of the left mandibular nerve. Upper part of mandibular ramus removed. *ATN* auriculotemporal nerve, *CT* chorda tympani, *FO* foramen ovale, *IAN* inferior alveolar nerve, *LN* lingual nerve, *MN* mandibular nerve

2014). In a series of cadavers, there was a 47% concordance in external morphology between the IAN and LN (Tan et al. 2014). Variations of the LN have been reported in the literature such as communicating branches with the IAN in up to 33% of cases (Rácz and Maros 1981). Additional communications have been noted as the nerve to mylohyoid, auriculotemporal nerve, and the hypoglossal nerve (Al-Amery et al. 2016; Piagkou et al. 2011; Păduraru and Rusu 2013). The LN may be injured during surgical procedures of the mouth leading to somatosensory disturbances including loss of taste (Al-Amery et al. 2016). Lingual nerve injury most commonly occurs during procedures involving the lower third molar due to its course being close in proximity (McGeachie 2002). The retraction of the lingual flap is purported to increase the incidence of LN injury during the third molar extraction (Baqain et al. 2010). When viewed histologically, the LN is more often a unifascicular nerve at the level of the lingula versus the polyfascicular IAN (Tan et al. 2014). It has been postulated that this may explain the increased incidence of LN versus IAN injury (Tan et al. 2014).

4.6 Sphenomandibular Ligament and Nerve to the Mylohyoid

The sphenomandibular ligament (SML) is a structure linking the spine of the sphenoid bone with the lingula of the mandible (Garg and Townsend 2001) (Fig. 4.8). The SML, due to its insertion upon the lingula of the mandible, is directly in the needle trajectory for an IAN blockade and, depending on the needle depth, may prevent adequate anesthetic distribution (Khoury et al. 2010; Garg and Townsend 2001). The degree to which the SML may interfere with anesthetic distribution may be rooted somewhat in the nature of its mandibular attachment. The SML attaches to the mandibular lingula but more often than not extends further to also insert in the mylohyoid groove (Shiozaki et al. 2007). A study of SML morphology demonstrated that over half of SMLs had a broad insertion from the lingula all the way to the posterior border of the mandibular ramus forming what appeared to be a septal-like structure in the PM (Shiozaki et al. 2007). One suggestion presented in the literature is for clinicians to position their needle tip superior to the lingula as to either avoid the anterior SML attachment entirely or to pierce the attachment to avoid the SML barrier to local anesthetic diffusion entirely (Garg and Townsend 2001). Interestingly, Simonds et al. (2017) reported a duplicated SML, which could potentially influence the procedure of the IAN blockade.

The nerve to the mylohyoid (NM) is classically portrayed as a motor nerve innervating the mylohyoid muscle and the anterior belly of the digastric muscle (Stein et al. 2007). The NM originates from the bifurcation of the mandibular nerve into the IAN and NM or from the IAN itself (Wolf et al. 2016). It was noted that historically the NM departs from the IAN approximately 5 mm proximal to the IAN's entrance into the mandibular foramen (Barker and Davies 1972). Subsequently, it pierces the SML and descends furthermore within the mylohyoid groove of the mandible (Barker and Davies 1972). The NM is now believed to carry sensory fibers

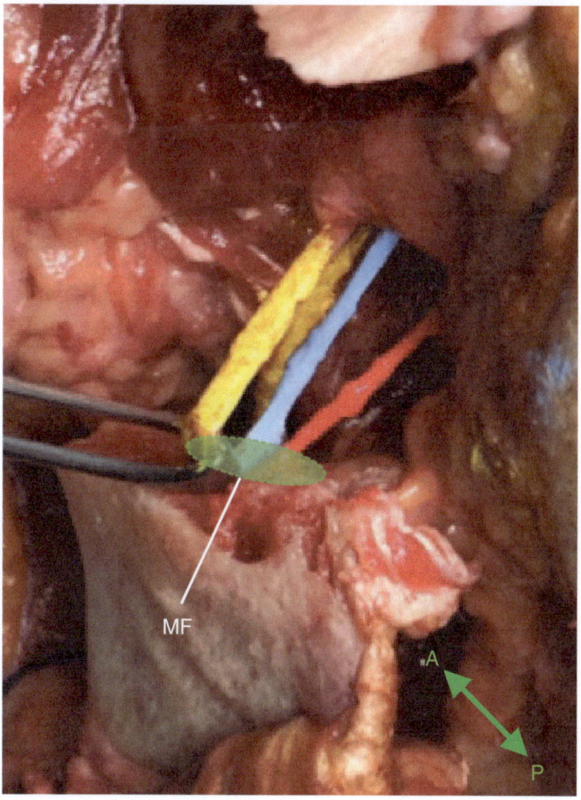

Fig. 4.8 Left sphenomandibular ligament (light blue). Note that the sphenomandibular ligament is located medial to the inferior alveolar nerve (yellow) and artery (red). *MF* mandibular foramen

for the mandibular teeth and may contribute to the high failure rate of an IAN blockade (Stein et al. 2007).

4.7 Mandibular Foramen and Lingula

The mandibular foramen (MF) is a canal located on the medial aspect of the mandibular ramus and provides an entrance for the inferior alveolar neurovascular bundle. The MF is bordered anterosuperiorly by a bony protuberance known as the lingula (Lipski et al. 2013b). The lingula is known to have variations of its own—some demonstrate the anterior/superior position, while some exhibit a medial location more reminiscent of the wall of the mandibular groove. A series of examinations conducted in 2012 demonstrate that the lingula is on average 5.82 mm below the lingula (Monnazzi et al. 2012). The MF opens into the mandibular canal at the top of the mylohyoid groove. On average the entrance to the MF has been shown to be 16.0 mm from the anterior border of the mandibular ramus and 10.2 mm from the posterior border (Sandhya et al. 2015). On a review of a CBCT series, it was demonstrated that approximately 46.5–65% of individuals possess accessory mandibular canals (Kalantar Motamedi et al. 2015).

4.8 Mylohyoid Groove

The mylohyoid groove (MHG) is a bony ridge which courses on the medial aspect of the mandible in an anteroinferior direction (Fig. 4.9). The MHG originates at the level of the MF and lingula and offers bony protection for the mylohyoid vessels and nerve, which course within the groove. It was demonstrated in 1979 that the MHG is open along its entire length in approximately 84% of cases (Arensburg and Nathan 1979). The presence of osseous deposition across the MHG is known as mylohyoid bridging and could be of concern for those performing procedures in the area (Nikolova et al. 2017).

4.9 Other Clinical Considerations

The PM and knowledge of its contents, boundaries, and variations are of vital importance to all clinicians performing surgical and anesthetic procedures in the region. It is now possible in certain cases to provide preoperative assessment of anatomical variants in the mandibular region with CT imaging techniques (Afsar et al. 1998). Routine dental procedures may not warrant such an investigation; however imaging knowledge of anatomic variants may abate some iatrogenic injury risk associated with certain operative undertakings. Numerous case reports have highlighted the complication of broken anesthetic needles retained within the PM (Bedrock et al. 1999; Catelani et al. 2013). Practitioner knowledge of PM anatomy is of the utmost importance to prevent further damage to locoregional structures during needle retrieval following these complications. In addition, since the PM is located close to the retromolar area, the inflammation or tumors, which appear in the retromolar area, could easily spread to the PM (Iwanaga et al. 2017b) (Fig. 4.10).

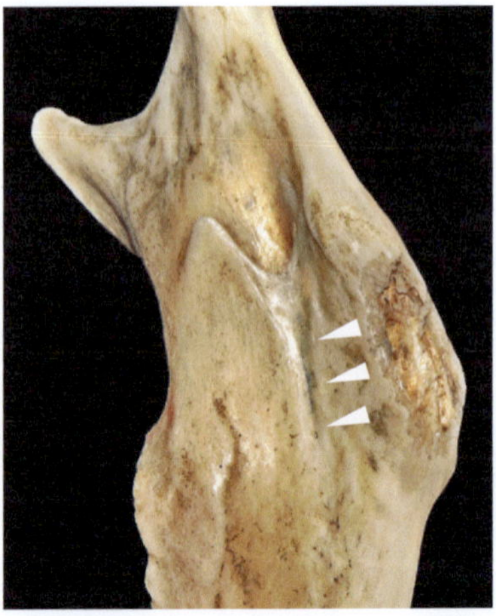

Fig. 4.9 Mylohyoid groove (arrowheads)

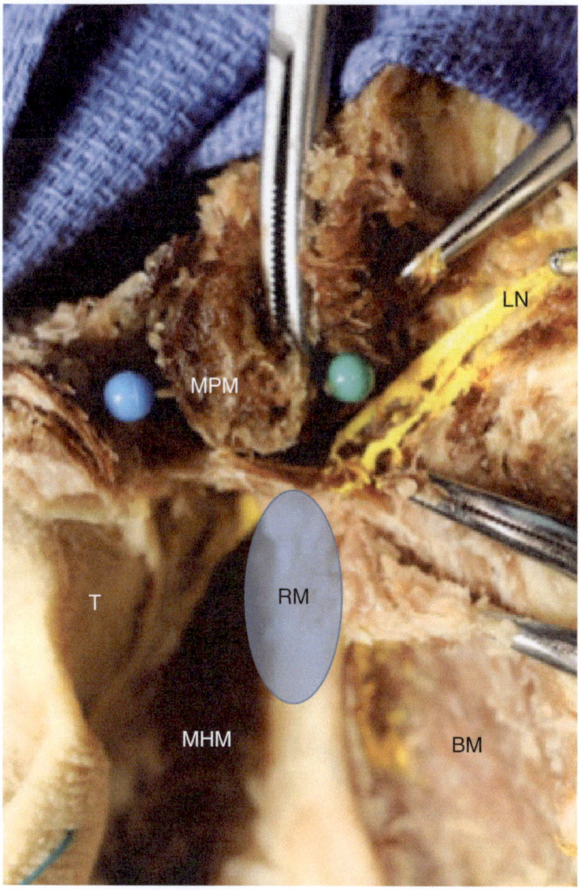

Fig. 4.10 Left pterygomandibular space (blue) and parapharyngeal space (green) (superior view). *BM* buccinator muscle, *LN* lingual nerve, *MHM* mylohyoid muscle, *MPM* medial pterygoid muscle, *RM* retromolar area, *T* tongue

References

Afsar A, Haas DA, Rossouw PE et al (1998) Radiographic localization of mandibular anesthesia landmarks. Oral Surg Oral Med Oral Pathol Oral Radiol Endod 86(2):234–241

Al-Amery SM, Nambiar P, Naidu M et al (2016) Variation in lingual nerve course: a human cadaveric study. PLoS One 11:e0162773

Anil A, Peker T, Turgut HB et al (2003) Variations in the anatomy of the inferior alveolar nerve. Br J Oral Maxillofac Surg 41:236–239

Arensburg B, Nathan H (1979) Anatomical observations on the mylohyoid groove, and the course of the mylohyoid nerve and vessels. J Oral Surg 37:93–96

Arias NH, Schejtman R, Devoto FC (1967) Distribution of components of the retromolar mandibular canal in the buccinator muscle. Rev Asoc Odontol Argent 55:502–503

Baqain ZH, Abukaraky A, Hassoneh Y et al (2010) Lingual nerve morbidity and mandibular third molar surgery: a prospective study. Med Princ Pract 19:28–32

Barker BCW, Davies PL (1972) The applied anatomy of the pterygomandibular space. Br J Oral Surg 10:43–55

Bedrock RD, Skigen A, Dolwick MF (1999) Retrieval of a broken needle in the pterygomandibular space. J Am Dent Assoc 130:685–687

Blaeser BF, August MA, Donoff RB et al (2003) Panoramic radiographic risk factors for inferior alveolar nerve injury after third molar extraction. J Oral Maxillofac Surg 61:417–421

Catelani C, Valente A, Rossi A et al (2013) Broken anesthetic needle in the pterygomandibular space. Four case reports. Minerva Stomatol 62:455–463

Frommer J, Mele FA, Monroe CW (1972) The possible role of the mylohyoid nerve in mandibular posterior tooth sensation. J Am Dent Assoc 85:113–117

Garg A, Townsend G (2001) Anatomical variation of the sphenomandibular ligament. Aust Endod J 27:22–24

Huelke D (1973) Selected dissections of the facial regions for advanced dental students, 6th edn. Overbeck Co., Ann Arbor, MI

Iwanaga J, Kikuta S, Nakamura M et al (2017) Intraoral vertico-sagittal ramus osteotomy: modification of the L-shaped osteotomy. Int J Oral Maxillofac Surg 46:1552–1556

Iwanaga J, Choi PJ, Vetter M et al (2018) Anatomical study of the lingual nerve and inferior alveolar nerve in the pterygomandibular space: complications of the inferior alveolar nerve block. Cureus 10:e3109

Jergenson MA, Norton NS, Opack JM et al (2005) Unique origin of the inferior alveolar artery. Clin Anat 18:597–601

Jorgensen N, Hayden J (1967) Premedication, local and general anesthesia in dentistry. Lea & Febiger, Philadelphia, PA

Kalantar Motamedi MH, Navi F, Sarabi N (2015) Bifid mandibular canals: prevalence and implications. J Oral Maxillofac Surg 73:387–390

Kaufman E, Weinstein P, Milgrom P (1984) Difficulties in achieving local anesthesia. J Am Dent Assoc 108:205–208

Khoury J, Mihailidis S, Ghabriel M et al (2010) Anatomical relationships within the human pterygomandibular space: relevance to local anesthesia. Clin Anat 23:936–944

Khoury JN, Mihailidis S, Ghabriel M et al (2011) Applied anatomy of the pterygomandibular space: improving the success of inferior alveolar nerve blocks. Aust Dent J 56:112–121

Lipski M, Lipska W, Motyl S et al (2013a) Anatomy of the pterygomandibular space - clinical implication and review. Folia Med Cracov 53:79–85

Lipski M, Tomaszewska IM, Lipska W et al (2013b) The mandible and its foramen: anatomy, anthropology, embryology and resulting clinical implications. Folia Morphol (Warsz) 72:285–292

Malamed S (2004) Handbook of local anesthesia, 5th edn. Mosby, St. Louis, MO

Matani JD, Kheur MG, Kheur SM et al (2014) The anatomic inter relationship of the neurovascular structures within the inferior alveolar canal: a cadaveric and histological study. J Maxillofac Oral Surg 13:499–502

McGeachie JK (2002) Anatomy of the lingual nerve in relation to possible damage during clinical procedures. Ann R Australas Coll Dent Surg 16:109–110

Monnazzi MS, Passeri LA, Gabrielli MF et al (2012) Anatomic study of the mandibular foramen, lingula and antilingula in dry mandibles, and its statistical relationship between the true lingula and the antilingula. Int J Oral Maxillofac Surg 41:74–78

Murphy TR, Grundy EM (1969) The inferior alveolar neurovascular bundle at the mandibular foramen. Dent Pract Dent Rec 20:41–48

Nakagawa Y, Ishii H, Nomura Y et al (2007) Third molar position: reliability of panoramic radiography. J Oral Maxillofac Surg 65:1303–1308

Nikolova SY, Toneva DH, Yordanov YA et al (2017) Morphometric study of the mylohyoid bridging in dry mandibles. Anthropol Anz 74:113–122

Păduraru D, Rusu MC (2013) The anatomy of the intralingual neural interconnections. Surg Radiol Anat 35:457–462

Piagkou M, Demesticha T, Skandalakis P et al (2011) Functional anatomy of the mandibular nerve: consequences of nerve injury and entrapment. Clin Anat 24:143–150

Rácz L, Maros T (1981) The anatomic variants of the lingual nerve in human (author's transl). Anat Anz 149:64–71

Sandhya K, Singh B, Lugun N et al (2015) Localization of mandibular foramen relative to landmarks in East Indian mandibles. Indian J Dent Res 26:571–575

Schejtman R, Devoto FC, Arias NH (1967) The origin and distribution of the elements of the human mandibular retromolar canal. Arch Oral Biol 12:1261–1268
Shaw MD, Fierst P (1988) Clinical prosection for dental gross anatomy: a medial approach to the pterygomandibular space. Anat Rec 222:305–308
Shields PW (1977) Further observations on mandibular anesthesia. Aust Dent J 22:334–337
Shiozaki H, Abe S, Tsumori N et al (2007) Macroscopic anatomy of the sphenomandibular ligament related to the inferior alveolar nerve block. Cranio 25:160–165
Sicher H, DuBrul E (1975) Oral anatomy, 6th edn. C.V. Mosby Company, St. Louis, MO
Simonds E, Iwanaga J, Oskouian RJ et al (2017) Duplication of the sphenomandibular ligament. Cureus 9:e1783
Stein P, Brueckner J, Milliner M (2007) Sensory innervation of mandibular teeth by the nerve to the mylohyoid: implications in local anesthesia. Clin Anat 20:591–595
Taghavi Zenouz A, Ebrahimi H, Mahdipour M et al (2008) The incidence of intravascular needle entrance during inferior alveolar nerve block injection. J Dent Res Dent Clin Dent Prospects 2:38–41
Takezawa K, Kageyama I (2015) Nerve fiber analysis on the morphology of the lingual nerve. Anat Sci Int 90:298–302
Tan VL, Andrawos A, Ghabriel MN et al (2014) Applied anatomy of the lingual nerve: relevance to dental anaesthesia. Arch Oral Biol 59:324–335
Traeger KA (1979) Hematoma following inferior alveolar injection: a possible cause for anesthesia failure. Anesth Prog 26:122–123
Wolf KT, Brokaw EJ, Bell A et al (2016) Variant inferior alveolar nerves and implications for local anesthesia. Anesth Prog 63:84–90

Anatomy and Variations of the Retromolar Fossa

5

Puhan He, Mindy K. Truong, and Shogo Kikuta

5.1 Introduction

The retromolar region is a triangular area bounded by the temporal crest on the medial side, anterior border of ramus on the lateral side, and the posterior portion of the third molar area. The deeper fibers of the temporalis muscle attach to the upper two-thirds of the medial boundary of the retromolar fossa. The superficial fibers of the temporalis muscle attach to the upper one-third of the lateral boundary of the retromolar triangle. The buccal nerve and artery cross the middle portion of the medial border, while the lingual nerve crosses the inferior portion. The retromolar foramen also consists of clinically relevant vessels and nerves (Truong et al. 2017). The posterior portion of the third molar area attaches to the pterygomandibular raphe, where the buccinator and superior pharyngeal constrictor muscle (SPCM) attach anteriorly and posteriorly (Horta et al. 2016). It is implicated clinically in procedures such as third molar extractions, periodontal distal wedge procedures, and sagittal split osteotomy. Furthermore, the retromolar triangle marks an aggressive type of oral cancer.

5.2 Anatomy and Contents of the Retromolar Fossa

5.2.1 Retromolar Fossa and Retromolar Triangle

The retromolar fossa is a shallow fossa located distally to the most posterior molar. It is bounded medially by the inferior portion of the temporal crest, laterally by the

P. He (✉) · M. K. Truong
Harvard School of Dental Medicine, Boston, MA, USA
e-mail: Puhan_He@hsdm.harvard.edu; Mindy_Truong@hsdm.harvard.edu

S. Kikuta
Kurume University School of Medicine, Kurume, Fukuoka, Japan

J. Iwanaga, R. S. Tubbs (eds.), *Anatomical Variations in Clinical Dentistry*,
https://doi.org/10.1007/978-3-319-97961-8_5

external oblique ridge, and anteriorly by the distal of the most posterior molar. The retromolar triangle is located in the retromolar fossa in the most inferior and anterior, and only horizontal, portion (Woelfel et al. 2012) (Fig. 5.1). The retromolar fossa may have clinical significance as a passing zone for infectious processes from the third molar region, which are called Chompret-L'Hirondel migratory abscesses (Galdames 2012).

The retromolar triangle is bordered by the two bifurcation lips of the distal surface of the last molar and the temporal ridge of the mandible. It is a permanent structure which does not rely on the presence of dentition, and its shape is not always triangular (Galdames 2012).

5.2.2 Retromolar Pad

The depressible mucosal elevation covering the retromolar triangle is called the retromolar pad or the piriformis papilla (Fig. 5.2). The retromolar pad has an antero-posteriorly elongated elevation that widens near the second molar. It has an average length of 11.2 mm and an average maximum transverse diameter of 7.9 mm. Its shape can be ovular (53.1%), rounded (29.6%), or triangular (17.3%) (Lopez et al.

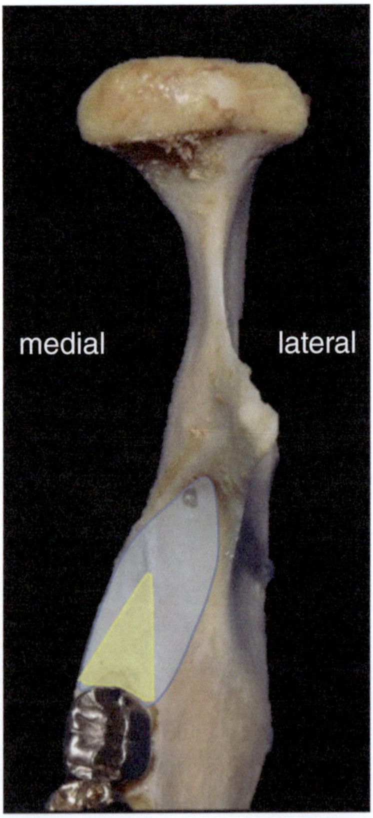

Fig. 5.1 Superior view of the retromolar fossa (blue) and retromolar triangle (yellow)

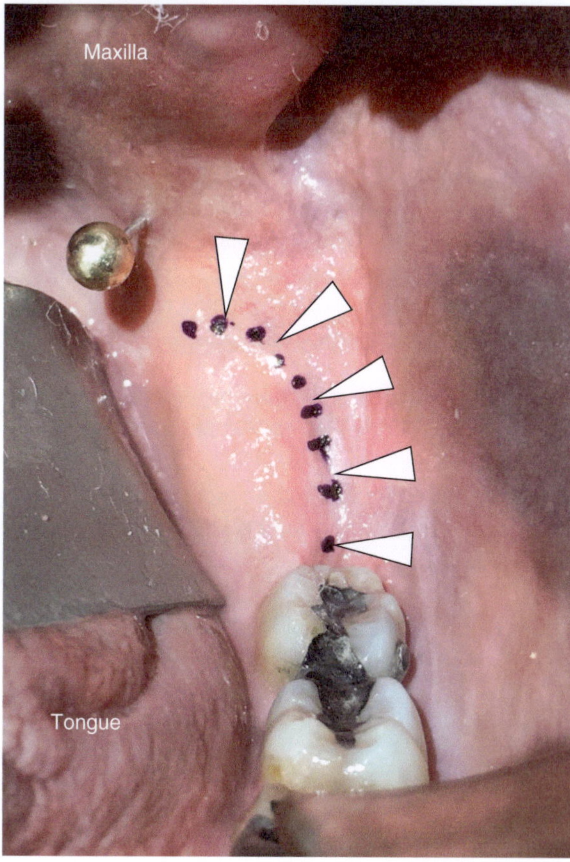

Fig. 5.2 Retromolar pad (arrowheads)

2008). Histologically, the lining of the retromolar pad corresponds to an epithelium with varying amounts of parakeratinization (Mccrorie and Hall 1965). The retromolar pad is clinically significant since it is considered a landmark for determination of the occlusal plane, and assists in posterior sealing, in the fabrication of a mandibular denture (Celebić et al. 1995; Taíeb and Carpentier 1989). Furthermore, because of its low vascularization and high capacity for stretching, the retromolar pad may be used as an alternative entry site for anesthesia of the inferior alveolar nerve (IAN), via the retromolar triangle, for patients who are carriers of blood dyscrasias (Suazo Galdames et al. 2008).

5.2.3 Musculature

Within the retromolar fossa area, the three main muscles are the buccinator, the SPCM, and the distal fascicle of the temporalis muscle. Additionally, the medial pterygoid and palatoglossus muscles may extend posterosuperiorly to the mandibular third molar area. The buccinator muscle fibers insert in the pterygomandibular raphe and are slanted inferolaterally toward its point of origin in the retromolar fossa. Its most posterior fibers attach within the retromolar triangle on the buccinator crest, a slight ridge of bone. The buccinator muscle and its fascia must be crossed during anesthetic technique toward the lingula. The SPCM crosses into the boundaries of the retromolar fossa area in two parts, the buccopharyngeal (pterygomandibular raphe) and mylopharyngeal (mylohyoid line of the mandible) parts. The distal fascicle of the temporalis muscle extends inferiorly along the temporal ridge and inserts in the medial lip of the retromolar triangle. This appears to play a role in the shape changes of the retromolar pad (Galdames 2012). Horta et al. (2016) observed that the superficial fibers of the temporalis muscle inserted into the upper third of the lateral boundary of the retromolar triangle. The deeper fibers of the temporalis muscle, arising from the anterior wall of the temporal fossa, attached to the upper two-thirds of the medial boundary of the retromolar triangle. The medial pterygoid muscle was observed deep to the medial border. Along the anterior border of the medial pterygoid and anteromedial to the triangle, the pterygomandibular raphe gave attachment to the buccinator muscle anteriorly and the SPCM posteriorly (Fig. 5.3).

5.2.4 Cortex and Cancellous Bone Tissue

The cortex and cancellous bone tissue of the retromolar triangle region has an average thickness of 1.77 mm. It widens toward the vestibule and thins toward the lingual (Suazo et al. 2008). The retromolar area is a possible site for autologous bone harvesting and is associated with lower rates of complication compared to sagittal

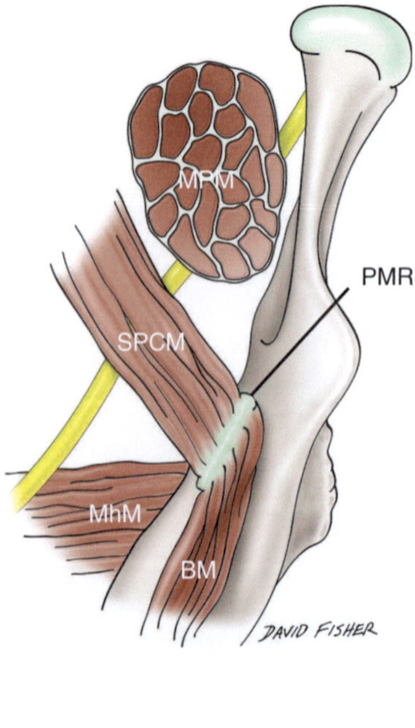

Fig. 5.3 Musculature in the retromolar region (superior view of the left mandible). *BM* buccinators, *MhM* mylohyoid muscle, *MPM* medial pterygoid muscle, *PMR* pterygomandibular raphe, *SPCM* superior pharyngeal constrictor muscle

split ramus osteotomy or distraction osteogenesis technique. However, care must be taken to avoid injuring nerves and vasculature in the region, including the mandibular canal (Di Bari et al. 2013, 2014).

5.2.5 Retromolar Foramen and Retromolar Canal

An anatomic variant called the retromolar foramen (RMF) may be located in the retromolar fossa (Fig. 5.4). The RMF is the bony hole through which the retromolar canal (RMC) exits in the mandible. The RMC has been studied by many researchers using cone beam computed tomography (CBCT), but there have been no studies observed from inside the mandibular canal. Iwanaga et al. (2017) showed the RMC branched from the mandibular canal using endoscopy. The RMC carries vasculature and innervation to structures in the area. Injury to the elements of the RMC during surgical procedures could lead to consequences such as bleeding, traumatic neuroma, pain, or paresthesia (Truong et al. 2017).

5.2.6 Vasculature and Lymphatics

The vasculature of the retromolar area is provided mainly by collateral branches of the inferior alveolar and buccal arteries, as well as the masseteric, ascending palatine, mylohyoid, sublingual, and submental arteries (Castelli et al. 1975; Kawai et al. 2006). The inferior alveolar artery begins at the first segment of the maxillary artery on its infratemporal path. It gives off the mylohyoid branch and then enters the mandible through the mandibular foramen. The inferior alveolar artery then runs through the mandibular canal to provide vasculature to the bone of the mandible, the posterior dentition (Galdames 2012).

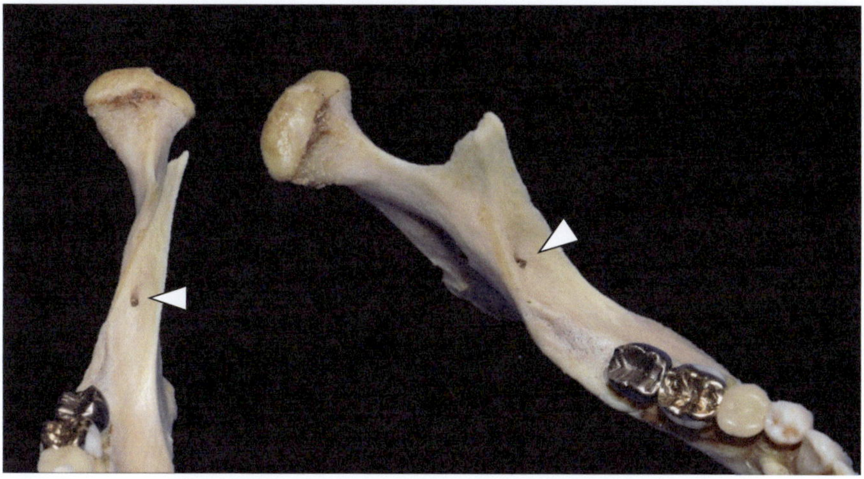

Fig. 5.4 Retromolar foramen (arrowheads) in the retromolar fossa

A case was reported which observed an artery which runs through the RMC, an anatomic variant present in some individuals. The artery branched from the inferior alveolar arteries and coursed through the RMC to join branches of the buccal and facial artery, before giving off the superior and inferior labial arteries (Kodera and Hashimoto 1995). Injury to this artery in the perimandibular retromolar region during surgical procedures may lead to excessive bleeding (Fukami et al. 2012). The buccal artery is a branch of the maxillary artery which provides vasculature to the buccinator muscle, as well as the region and mucosa of the posterior mandible. The buccal artery provides collateral vasculature which exits the mandible through the RMF. The veins of the retromolar area follow the course taken by the arteries. Most blood is directed to the pterygoid venous plexus (Galdames 2012). The vasculature which runs through the RMC may spread infection or tumors from the oropharynx to systemic circulation (Potu et al. 2014a).

The lymph of the retromolar area drains mainly into the superior-deep jugular lymph nodes, and there may be some drainage into the retropharyngeal and periparotid nodes. Lymphatic drainage is clinically significant in the early spread of oral carcinoma (Horta et al. 2016).

5.2.7 Course of the Buccal Nerve

The buccal nerve originates from the anterior division of the mandibular nerve (CN V_3), one of the three branches of the trigeminal nerve. The buccal nerve provides sensory innervation to the posterior mandibular buccal gingiva, the buccal mucosa, the skin around the angle of the mouth, and the mucosa of the lateral parts of the upper and lower lips. From its origin, the buccal nerve courses between the two heads of the lateral pterygoid muscle, under the inferior portion of the temporalis muscle, and medial to the mandible. As the buccal nerve approaches the retromolar region, it pierces the posterior half of the buccinator muscle to arborize extensively on the buccal surface of the cheek (Tubbs et al. 2010). The buccinator originates from the pterygomandibular raphe, a component of the retromolar fossa. To perform a buccal nerve block, the injection should be placed immediately prior to the location where the buccal nerve branches into its posterior buccal mucosal branches. This is at the anterior distal insertion of the tendon of the temporalis, along the apex of the retromolar triangle (Benninger and Lee 2012; Takezawa et al. 2017).

5.2.8 Course of the Lingual Nerve

The lingual nerve originates from the posterior division of the mandibular nerve. It supplies general sensation to the mucosa of the anterior two-thirds of the tongue, the sublingual mucosa, the mandibular lingual gingiva, and the floor of the mouth, as well as parasympathetic innervation for the sublingual and submandibular glands. The lingual nerve descends with the IAN on the medial aspect of the lateral pterygoid muscle. The chorda tympani joins the lingual nerve as it continues inferiorly. The lingual nerve enters the mouth by passing beneath the lower border of the

SPCM (Fig. 5.5). It courses along the periosteum on the medial surface of the mandible and traverses the region below the root of the lower third molar.

Here, it is only covered by the gingival mucoperiosteum. The lingual nerve has variable locations in relation to the retromolar fossa. In 57.4–62% the lingual nerve is in immediate contact with the lingual cortex medial to the retromolar triangle (Hölzle and Wolff 2001; Kiesselbach and Chamberlain 1984) (Fig. 5.6). The lingual nerve has been found 17.6% of the time at the level of the alveolar crest or higher in this region (Kiesselbach and Chamberlain 1984). This means that one in six patients will have either their left or right lingual nerve located above the lingual plate in the third molar region. Less often, the lingual nerve is found in the retromolar pad region at 0.15% and 1.5% of the cases (Behnia et al. 2000; Kiesselbach and Chamberlain 1984). Notably, the nerve may be within 1 mm of the bone on the lingual or distal side of the third molar with a range from 1 to 7 mm (Pogrel et al. 1995).

Thereafter, the lingual nerve leaves the retromolar region and courses anteriorly just superior to the mylohyoid muscle within the floor of the mouth. Seventy-five percent of lingual nerves turned toward the tongue at the first and second molar region (Chan et al. 2010). The lingual nerve travels between the sublingual and submandibular glands. In its course, the lingual nerve establishes communications with the mylohyoid nerve, which completes its innervation (Al-Amery et al. 2016). The course of the lingual nerve is also affected by the position of the tongue (Iwanaga 2017). Understanding the dynamic course of the lingual nerve could mitigate injury during surgery as well as improve local anesthesia success. It also highlights the retromolar region as a vulnerable location for the lingual nerve as it passes medially to the mandibular third molar (Sittitavornwong et al. 2017).

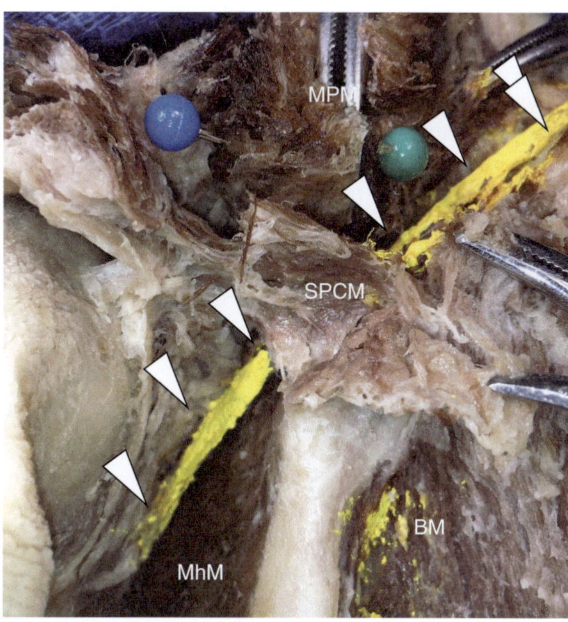

Fig. 5.5 Superior view of the course of the lingual nerve (arrowheads). *BM* buccinator muscle, *MhM* mylohyoid muscle, *MPM* medial pterygoid muscle, *SPCM* superior pharyngeal constrictor muscle

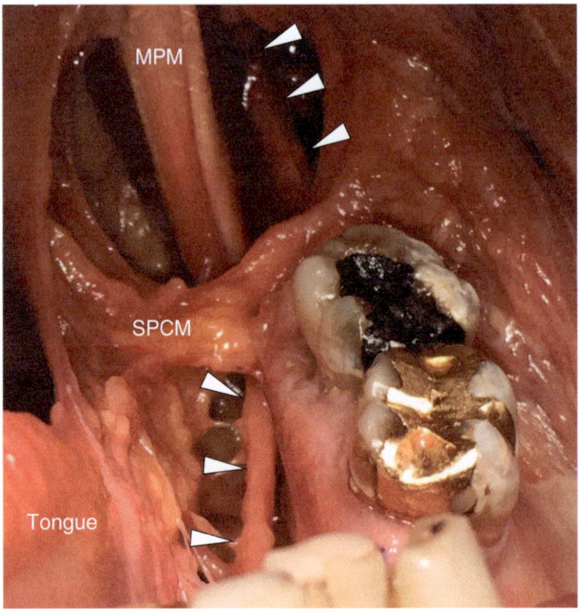

Fig. 5.6 The lingual nerve (arrowheads) in immediate contact with the lingual cortex medial to the retromolar triangle. *MPM* medial pterygoid muscle, *SPCM* superior pharyngeal constrictor muscle

5.3 Retromolar Foramen and Canal

RMF are apertures of the RMC in the mandible, located posteriorly to the last molar in the retromolar triangle. RMF are located mostly in the medial aspect of the retromolar fossa, proximal to the lingula (Potu et al. 2014a). The RMC is significant due to the presence of a neurovascular bundle which runs through it (Alves and Deana 2015).

The RMC is classified as a type 1 bifidity (unilateral/bilateral transverse bifidity) of the mandibular canal (Motamedi et al. 2016), (Kalantar Motamedi et al. 2015). A bifid mandibular canal (BMC) is an anatomic variant in which the mandibular canal divides into two parts, and each branch may carry its own neurovascular bundle (Kang et al. 2014). It has been suggested that bifid and trifid mandibular canals originate from incomplete fusion of separate mandibular canal nerves of the incisors and primary/permanent molars during embryonic development (Chávez-Lomeli et al. 1996). It is suggested that the nerve that runs through the RMC may arise from early accessory branches of the IAN or buccal nerve (Han and Hwang 2014).

5.3.1 Classifications of RMC Course

The multiple possible courses of the RMC have been categorized into numerous classifications. Ossenberg (1987) classified the RMC into three types (Fig. 5.7). A type 1 RMC has a vertically curved course and branches from the mandibular canal near the molars from a single mandibular foramen. A type 2 RMC takes a horizontally curved course and branches from the mandibular canal just beyond the single

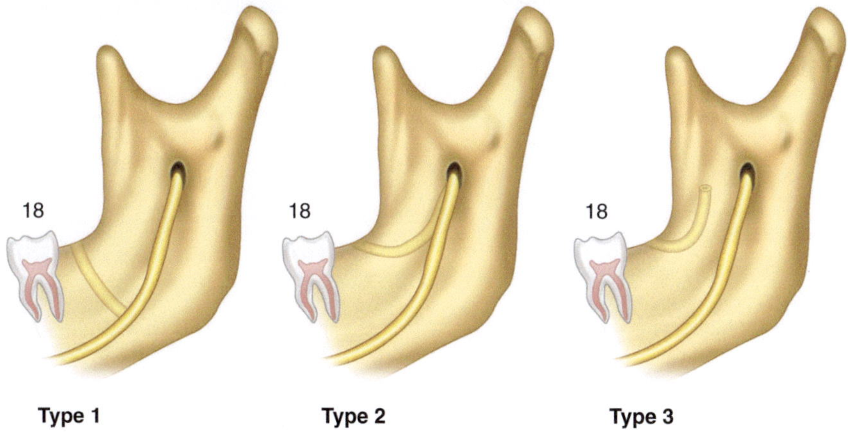

Fig. 5.7 Ossenberg's classification

mandibular foramen. A type 3 RMC has a separate foramen in the mandibular ramus and runs independently of the mandibular canal. It is thought that the type 3 RMC arises in the infratemporal fossa as a terminal branch of the anterior trunk of the IAN and courses forward, terminating in the buccal gingiva, mucosa, and skin. It usually passes medial to the most inferior part of the temporalis tendon at its insertion on the coronoid process. The type 3 RMC is assumed to convey all or a portion of the buccal nerve and is referred to as temporal crest canal by some authors (Han and Park 2013).

5.3.2 Contents of the RMC

The RMC originates from the mandibular canal, follows a recurrent path, and exits the mandible in RMF or nearby foramen. The contents of the RMC are derived from their inferior dental homologues, including a myelinated nerve (Fig. 5.8), and one or more arterioles and venules (Schejtman et al. 1967). Within the RMC, an artery is present with a lumen of 120–130 μm (Bilecenoglu and Tuncer 2006). After exiting the body of the mandible, the contents of the neurovascular bundle distribute mainly upon the temporalis tendon, the buccinator muscle, the most posterior portion of the alveolar process, and the mandibular third molar (Schejtman et al. 1967). The RMC may provide accessory innervation to the mandibular molars from the mandibular canal. Furthermore the RMC, particularly types 2 and 3, may contain aberrant buccal nerves which penetrate the buccinator muscle (Han and Hwang 2014).

5.3.3 Anesthesia

The nerve fibers which run through the RMC may prevent full anesthesia from conventional methods such as the IAN block. These nerves of the RMC may innervate

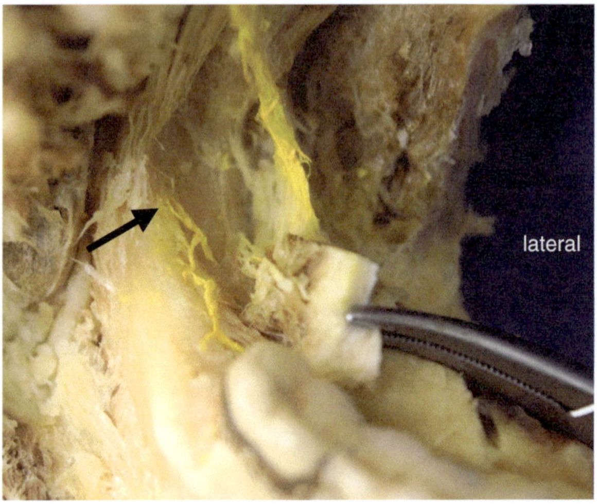

Fig. 5.8 Retromolar nerve arising from the retromolar foramen (arrow)

the third molars, buccal gingiva of mandibular molars and premolars, buccinator muscle, temporal tendon, and posterior portion of the mandible (Bilecenoglu and Tuncer 2006; Kodera and Hashimoto 1995; Schejtman et al. 1967). In this case, a few drops of anesthesia injected into the retromolar region may achieve complete anesthesia (Filo et al. 2015). Alternative techniques include the Gow-Gates and Akinosi-Vazirani techniques, which are indicated for any mandibular procedures, and especially when the patient has a history of standard IAN block failure due to accessory innervation or anatomic variation (Haas 2011; Pyle et al. 1999).

5.3.4 Vasculature of the RMC

The artery which runs through the RMC bifurcates into facial and buccal branches after exiting the RMF. Injury to this artery in the perimandibular retromolar region during surgical procedures may lead to excessive bleeding (Fukami et al. 2012). The vasculature which runs through the RMC may spread infection or tumors from the oropharynx to systemic circulation (Potu et al. 2014a).

5.3.5 Radiographic Detection

Several forms of radiography may be used to detect RMF and RMC, such as cone beam computed tomography (CBCT) (Fig. 5.9), computed tomography (CT), and panoramic radiography (PAN). Multiple studies have shown CBCT to be the most sensitive technique in detecting these anatomic variants compared to PAN and CT (Fukami et al. 2012; Han and Park 2013; Muinelo-Lorenzo et al. 2014; Naitoh et al. 2009; Sisman et al. 2015; von Arx et al. 2011). Motamedi et al. (2015) reported a prevalence of the RMC of less than 1% when utilizing PAN, while CBCT studies have detected a much higher incidence. CBCT can be used when a preliminary radiograph, such as a PAN, cannot provide clear definition of anatomical structures

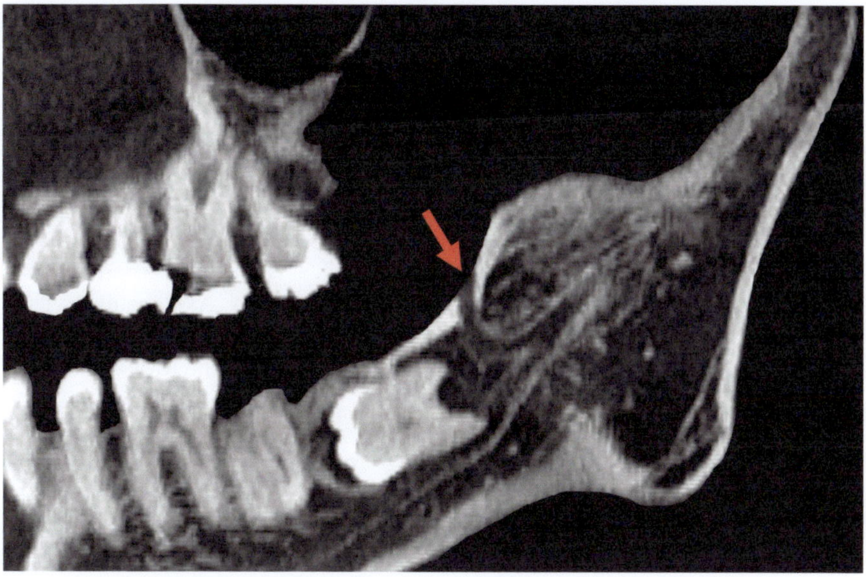

Fig. 5.9 Retromolar foramen (arrow) and canal detected by CBCT

in the mandibular molar area (Lizio et al. 2013). Limited CBCT may also be extremely valuable for the assessment of the RMF and RMC. It is important to localize anatomic variants such as the RMC prior to dentoalveolar surgery which may impact the retromolar area. This is especially significant when the presence of BMCs is suspected by PAN (Fukami et al. 2012).

5.3.6 Frequency

There is wide variation in the frequency, location, diameter, and distance of the canal in different individuals. The reported frequency of the RMF can range widely based on the visualization technology utilized and the geographic area studied. CBCT studies have reported a range from 5.4% (Kang et al. 2014) to 75.4% (Patil et al. 2013). PAN studies report a range from 3.06% (Sisman et al. 2015) to 8.8% (Capote et al. 2015). Human dry mandible studies report a range from 3.2% (Ossenberg 1987) to 72% (Schejtman et al. 1967). This wide range can be attributed to several factors including environmental and genetic factors, ethnic differences, and sample size variation across studies (Haas et al. 2016). However, studies have suggested that RMF and RMC are normal anatomical variations of the IAN, rather than anomalies (Motamedi et al. 2016).

5.3.7 Diameter and Location

Studies have reported the diameter of RMF to range from 0.2 mm (Ogawa et al. 2016) to 3.29 mm (Kang et al. 2014). Males on average have slightly larger diameters of RMF than females, which can be explained by the larger size of their mandibles (Motamedi et al. 2016). Reports of the mean distance between the RMF and distal edge of the third molar ranged from 4.23 mm (Bilecenoglu and Tuncer 2006) to 10.5 mm (Schejtman et al. 1967). The reported mean distances between the RMF and the distal edge of the second molar ranged from 11.91 mm (Bilecenoglu and Tuncer 2006) to 16.8 mm (Gamieldien and Van Schoor 2016). This suggests that the location of the RMF is not constant between individuals. To reveal how the RMC and RMF could affect mandibular anatomy, Kikuta et al. (2018) examined for the distance from the mandibular foramen to the distal end of the second molar in patients with RMC and patients without RMC, respectively. As a result, patients with RMC had significantly the longer distance than patients without RMC. This finding may help to understand the biological development of the RMC.

5.3.8 Age, Gender, Side Predilection, and Number

According to Ossenberg (1987), the peak incidence of RMF occurs in adolescents. This may be due to their increased neurovascular requirements for growth and eruption of the third molars. The preferential distribution of the nerve of the RMC on the temporalis tendon may relate to the adolescent peak of RMF and increased masticatory strength. No studies have reported a gender predilection for the presence of RMF. Gamieldien and Van Schoor (2016) state that sidedness of the RMF is not likely to have any developmental, surgical, or anatomical significance. Alves et al. (2015) reported a case of bilateral double RMF and a case of double left RMF. He et al. (2017) reported the only known case of a tripled RMF.

5.4 Clinical Application

5.4.1 Extraction of Third Molar and Infections

The retromolar fossa, located 4 to 11 mm from the third molar, is a clinically important consideration for extraction of third molars (Potu et al. 2014b). Damage to the blood vessels traversing through the RMF could partially explain intraoperative bleeding or postoperative hematomas in third molar extractions (Fig. 5.10). This close relation of the RMF and third molar could lead to damage of the structures traversing the RMF during third molar extractions and could be a reason for postoperative hematomas due to rupturing of the vessels in RMF (Suazo Galdames 2012). Another complication is paresthesia following the innervation of the buccal nerve which sometimes crosses the RMF (Singh 1981). Kikuta et al. (2018) revealed that one case had postoperative nerve paralysis of the buccal gingiva in the molar region in the retrospective study using CBCT of their facility. In this study, there was no unexpected bleeding during extraction of third molars. Other dysfunctions due to injury to the RMF include that of the temporalis and buccinator muscles (Potu et al. 2014b).

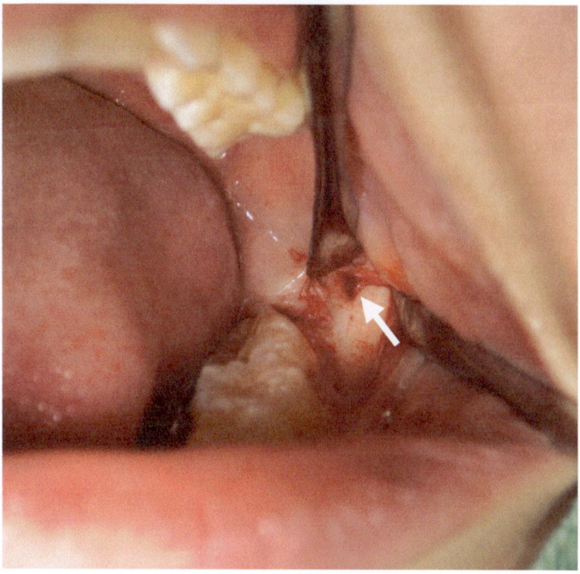

Fig. 5.10 Retromolar foramen (arrow) detected

5.4.2 Sagittal Split Ramus Osteotomy

Obwegeser developed the sagittal split ramus osteotomy (SSRO) in 1957 to avoid skin incision and create broad contacting bone surfaces. A bilateral SSRO would then divide the mandible into two smaller condyle-bearing segments and a larger segment consisting of the mandibular body including the teeth and chin. The condyle-bearing segment and the mandibular body are then fixated with either bicortical screws or mini-plate on either side of the osteotomy (Obwegeser 2007).

While postoperative IAN disturbances are common following SSRO ranging from 20 to 85% (He et al. 2017), with the popularization of screw fixation, the incidence of lingual nerve injury has declined. Current cases of lingual nerve injury after SSRO were primarily due to bicortical screw placement near the superior border of the mandible in the region of the third molar and retromolar region (Monson 2013). It is vital to understand the variable locations that the lingual nerve may traverse the retromolar region.

5.4.3 Retromolar Triangle Tumors

The retromolar triangle (RMT) is a rare but aggressive site for oral cancer. While the most frequent type of RMT tumor is squamous cell carcinoma, a case of clear cell carcinoma (Yamamoto et al., 2014) and of leiomyoma (Luaces Rey et al. 2007) have been reported. RMT tumors are aggressive because of their spatial relationship to a multitude of structures. Particularly, RMT has a predilection for bony invasion as well as perineural and lymphatic spread. Bony invasion of the mandible is common because there is little soft tissue coverage in the trigone. Soft tissue involvement often calls for MRI to detect perineural spread into regional nerves such as the buccal and lingual nerves (Horta et al. 2016).

The nature of the tumor requires an aggressive treatment approach requiring either intraoral or extraoral surgical resection and sometimes mandibulectomy. The treatment can also cause debilitating complications because of damage to regional structures such as the lingual nerve causing neurosensory defect, as well as muscles of speech and swallowing (Horta et al. 2016).

5.4.4 Distal Wedge Procedure

Deep periodontal pockets associated with the distal surfaces of last molars in the maxilla and mandible are often seen because of difficulties in maintaining hygiene in this area. The retromolar area has minimal keratinized tissue, and the tissue is often mucosal glandular tissue; thus gingivectomy cannot be used to treat these deep probing depths and improve cleansibility (Cohen 2007). In 1963, Robinson described the "distal wedge operation" to manage to this condition. The wedge designs include triangular, square, and linear. The triangular incision is the most commonly used; however, this option requires sufficient keratinized tissue. The distal wedge procedure allows for increased access to treatment of osseous tissue and healing by primary intention, both improving outcomes (Robinson 1966).

References

Al-Amery SM, Nambiar P, Naidu M et al (2016) Variation in lingual nerve course: a human cadaveric study. PLoS One 11:e0162773. https://doi.org/10.1371/journal.pone.0162773

Alves N, Deana NF (2015) Anatomical and radiographical study of the retromolar canal and retromolar foramen in macerated mandibles. Int J Clin Exp Med 8:4292–4296

Behnia H, Kheradvar A, Shahrokhi M (2000) An anatomic study of the lingual nerve in the third molar region. J Oral Maxillofac Surg 58:649–651. –653

Benninger B, Lee B-I (2012) Clinical importance of morphology and nomenclature of distal attachment of temporalis tendon. J Oral Maxillofac Surg 70:557–561. https://doi.org/10.1016/j.joms.2011.02.047

Bilecenoglu B, Tuncer N (2006) Clinical and anatomical study of retromolar foramen and canal. J Oral Maxillofac Surg 64:1493–1497. https://doi.org/10.1016/j.joms.2006.05.043

Capote TS d O, Gonçalves M d A, Campos JÁDB (2015) Retromolar canal associated with age, side, sex, bifid mandibular canal, and accessory mental foramen in panoramic radiographs of Brazilians. Anat Res Int 2015:434083. https://doi.org/10.1155/2015/434083

Castelli WA, Nasjleti CE, Díaz-Pérez R (1975) Interruption of the arterial inferior alveolar flow and its effects on mandibular collateral circulation and dental tissues. J Dent Res 54:708–715. https://doi.org/10.1177/00220345750540040301

Celebić A, Valentić-Peruzović M, Kraljević K et al (1995) A study of the occlusal plane orientation by intra-oral method (retromolar pad). J Oral Rehabil 22:233–236

Chan H-L, Leong DJM, Fu J-H et al (2010) The significance of the lingual nerve during periodontal/implant surgery. J Periodontol 81:372–377. https://doi.org/10.1902/jop.2009.090506

Chávez-Lomeli ME, Mansilla Lory J, Pompa JA et al (1996) The human mandibular canal arises from three separate canals innervating different tooth groups. J Dent Res 75:1540–1544. https://doi.org/10.1177/00220345960750080401

Cohen ES (2007) Atlas of cosmetic and reconstructive periodontal surgery. PMPH-USA
Di Bari R, Coronelli R, Cicconetti A (2013) Radiographic evaluation of the symphysis menti as a donor site for an autologous bone graft in pre-implant surgery. Imaging Sci Dent 43:135–143. https://doi.org/10.5624/isd.2013.43.3.135
Di Bari R, Coronelli R, Cicconetti A (2014) An anatomical radiographic evaluation of the posterior portion of the mandible in relation to autologous bone harvest procedures. J Craniofac Surg 25:e475–e483. https://doi.org/10.1097/SCS.0000000000000598
Filo K, Schneider T, Kruse AL et al (2015) Frequency and anatomy of the retromolar canal—implications for the dental practice. Swiss Dent J 125:278–292
Fukami K, Shiozaki K, Mishima A et al (2012) Bifid mandibular canal: confirmation of limited cone beam CT findings by gross anatomical and histological investigations. Dento Maxillo Facial Radiol 41:460–465. https://doi.org/10.1259/dmfr/60245722
Galdames I (2012) Lower Third Molar Region [WWW Document]. URL/paper/Lower-Third-Molar-Region-Tercer-Galdames/279fe6c8550ee7b53f12472968a5f28b4d0f29d9. Accessed 20 Nov 17
Gamieldien MY, Van Schoor A (2016) Retromolar foramen: an anatomical study with clinical considerations. Br J Oral Maxillofac Surg 54:784–787. https://doi.org/10.1016/j.bjoms.2016.05.011
Haas DA (2011) Alternative mandibular nerve block techniques: a review of the Gow-Gates and Akinosi-Vazirani closed-mouth mandibular nerve block techniques. J Am Dent Assoc 142:8S–12S
Haas LF, Dutra K, Porporatti AL et al (2016) Anatomical variations of mandibular canal detected by panoramic radiography and CT: a systematic review and meta-analysis. Dento Maxillo Facial Radiol 45:20150310. https://doi.org/10.1259/dmfr.20150310
Han S-S, Hwang Y-S (2014) Cone beam CT findings of retromolar canals in a Korean population. Surg Radiol Anat SRA 36:871–876. https://doi.org/10.1007/s00276-014-1262-1
Han S-S, Park C-S (2013) Cone beam CT findings of retromolar canals: report of cases and literature review. Imaging Sci Dent 43:309–312. https://doi.org/10.5624/isd.2013.43.4.309
He P, Iwanaga J, Matsushita Y et al (2017) A comparative review of mandibular orthognathic surgeries with a focus on intraoral Vertico-sagittal ramus osteotomy. Cureus 9
He P, Iwanaga J, Truong MK et al (2017) First report of tripled retromolar foramina. Cureus 9:e1440. https://doi.org/10.7759/cureus.1440
Hölzle FW, Wolff KD (2001) Anatomic position of the lingual nerve in the mandibular third molar region with special consideration of an atrophied mandibular crest: an anatomical study. Int J Oral Maxillofac Surg 30:333–338. https://doi.org/10.1054/ijom.2001.0064
Horta R, Nascimento R, Silva A et al (2016) The Retromolar Trigone: anatomy, cancer treatment modalities, reconstruction, and a classification system. J Craniofac Surg 27:1070–1076. https://doi.org/10.1097/SCS.0000000000002625
Iwanaga J (2017) The clinical view for dissection of the lingual nerve with application to minimizing iatrogenic injury. Clin Anat N Y N 30:467–469. https://doi.org/10.1002/ca.22877
Iwanaga J, Watanabe K, Saga T et al (2017) A novel method for observation of the mandibular foramen: application to a better understanding of dental anatomy. Anat Rec 300:1875–1880
Kalantar Motamedi MH, Navi F, Sarabi N (2015) Bifid mandibular canals: prevalence and implications. J Oral Maxillofac Surg 73:387–390. https://doi.org/10.1016/j.joms.2014.09.011
Kang J-H, Lee K-S, Oh M-G et al (2014) The incidence and configuration of the bifid mandibular canal in Koreans by using cone-beam computed tomography. Imaging Sci Dent 44:53–60. https://doi.org/10.5624/isd.2014.44.1.53
Kawai T, Sato I, Yosue T et al (2006) Anastomosis between the inferior alveolar artery branches and submental artery in human mandible. Surg Radiol Anat SRA 28:308–310. https://doi.org/10.1007/s00276-006-0097-9
Kiesselbach JE, Chamberlain JG (1984) Clinical and anatomic observations on the relationship of the lingual nerve to the mandibular third molar region. J Oral Maxillofac Surg 42:565–567
Kikuta S, Iwanaga J, Nakamura K et al (2018) The retromolar canals and foramina: radiographic observation and application to oral surgery. Surg Radiol Anat 40:647–652. (submitted to J Craniofac Surg)

Kodera H, Hashimoto I (1995) A case of mandibular retromolar canal: elements of nerves and arteries in this canal. Kaibogaku Zasshi 70:23–30

Lizio G, Pelliccioni GA, Ghigi G et al (2013) Radiographic assessment of the mandibular retromolar canal using cone-beam computed tomography. Acta Odontol Scand 71:650–655. https://doi.org/10.3109/00016357.2012.704393

Lopez FB, Cantin LM, Sandoval MC (2008) Biomethrics study of the retromolar pad. Int J Odontostomat 2:39–42

Luaces Rey R, Lorenzo Franco F, Gómez Oliveira G et al (2007) Oral leiomyoma in retromolar trigone: A case report. Medicina Oral, Patología Oral y Cirugía Bucal (Internet), 12:53–55

Mccrorie JW, Hall DC (1965) A histological investigation of the retromolar and pear-shaped pad. Dent Pract Dent Rec 15:237–239

Monson LA (2013) Bilateral sagittal split osteotomy. Semin Plast Surg 27:145–148. https://doi.org/10.1055/s-0033-1357111

Motamedi MHK, Gharedaghi J, Mehralizadeh S et al (2016) Anthropomorphic assessment of the retromolar foramen and retromolar nerve: anomaly or variation of normal anatomy? Int J Oral Maxillofac Surg 45:241–244. https://doi.org/10.1016/j.ijom.2015.10.017

Muinelo-Lorenzo J, Suárez-Quintanilla JA, Fernández-Alonso A et al (2014) Descriptive study of the bifid mandibular canals and retromolar foramina: cone beam CT vs panoramic radiography. Dento Maxillo Facial Radiol 43:20140090. https://doi.org/10.1259/dmfr.20140090

Naitoh M, Hiraiwa Y, Aimiya H et al (2009) Observation of bifid mandibular canal using cone-beam computerized tomography. Int J Oral Maxillofac Implants 24:155–159

Obwegeser HL (2007) Orthognathic surgery and a tale of how three procedures came to be: a letter to the next generations of surgeons. Clin Plast Surg 34:331–355. https://doi.org/10.1016/j.cps.2007.05.014

Ogawa A, Fukuta Y, Nakasato H et al (2016) Evaluation by dental cone-beam computed tomography of the incidence and sites of branches of the inferior dental canal that supply mandibular third molars. Br J Oral Maxillofac Surg. https://doi.org/10.1016/j.bjoms.2016.08.007

Ossenberg NS (1987) Retromolar foramen of the human mandible. Am J Phys Anthropol 73:119–128. https://doi.org/10.1002/ajpa.1330730112

Patil S, Matsuda Y, Nakajima K et al (2013) Retromolar canals as observed on cone-beam computed tomography: their incidence, course, and characteristics. Oral Surg Oral Med Oral Pathol Oral Radiol 115:692–699. https://doi.org/10.1016/j.oooo.2013.02.012

Pogrel MA, Renaut A, Schmidt B et al (1995) The relationship of the lingual nerve to the mandibular third molar region: an anatomic study. J Oral Maxillofac Surg 53:1178–1181. https://doi.org/10.1016/0278-2391(95)90630-4

Potu BK, Kumar V, Salem A-H et al (2014a) Occurrence of the retromolar foramen in dry mandibles of South-eastern part of India: a morphological study with review of the literature. Anat Res Int 2014:296717. https://doi.org/10.1155/2014/296717

Potu BK, Kumar V, Salem A-H, et al (2014b) Occurrence of the retromolar foramen in dry mandibles of south-eastern part of India: a morphological study with review of the literature [WWW Document]. Anat Res Int. https://www.hindawi.com/journals/ari/2014/296717/. Accessed 10 Dec 17

Pyle MA, Jasinevicius TR, Lalumandier JA et al (1999) Prevalence and implications of accessory retromolar foramina in clinical dentistry. Gen Dent 47:500–503. 505

Robinson RE (1966) The distal wedge operation. Periodontics 4:256–264

Schejtman R, Devoto FC, Arias NH (1967) The origin and distribution of the elements of the human mandibular retromolar canal. Arch Oral Biol 12:1261–1268

Singh S (1981) Aberrant buccal nerve encountered at third molar surgery. Oral Surg Oral Med Oral Pathol 52:142. https://doi.org/10.1016/0030-4220(81)90310-8

Sisman Y, Ercan-Sekerci A, Payveren-Arikan M et al (2015) Diagnostic accuracy of cone-beam CT compared with panoramic images in predicting retromolar canal during extraction of impacted mandibular third molars. Med Oral Patol Oral Cir Bucal 20:e74–e81

Sittitavornwong S, Babston M, Denson D et al (2017) Clinical anatomy of the lingual nerve: a review. J Oral Maxillofac Surg 75:926.e1–926.e9. https://doi.org/10.1016/j.joms.2017.01.009

Suazo Galdames I (2012) Lower third molar region. Int J Morphol 30:970–978. https://doi.org/10.4067/S0717-95022012000300034

Suazo Galdames IC, Cantín López MG, Zavando Matamala DA (2008) Inferior alveolar nerve block anesthesia via the retromolar triangle, an alternative for patients with blood dyscrasias. Med Oral Patol Oral Cirugia Bucal 13:E43–E47

Suazo GIC, Zavando MDA, Cantin LM (2008) Retromolar canal and foramen prevalence in dried mandibles and clinical implications. Int J Odontostomat 2:183–187

Taíeb F, Carpentier P (1989) Anatomy of the maxillary and mandibular retromolar area: effect on complete dentures. 2. The mandibular retromolar region. Cah Prothese 67:112–119

Takezawa K, Ghabriel M, Townsend G (2017) The course and distribution of the buccal nerve: clinical relevance in dentistry. Aust Dent J 63:66–71. https://doi.org/10.1111/adj.12543

Truong MK, He P, Adeeb N et al (2017) Clinical anatomy and significance of the retromolar foramina and their canals: a literature review. Cureus 9:e1781

Tubbs RS, Johnson PC, Loukas M et al (2010) Anatomical landmarks for localizing the buccal branch of the trigeminal nerve on the face. Surg Radiol Anat 32:933–935. https://doi.org/10.1007/s00276-010-0656-y

von Arx T, Hänni A, Sendi P et al (2011) Radiographic study of the mandibular retromolar canal: an anatomic structure with clinical importance. J Endod 37:1630–1635. https://doi.org/10.1016/j.joen.2011.09.007

Woelfel JB, Scheid RC, Weiss G (2012) Woelfel's dental anatomy. Wolters Kluwer/Lippincott Williams & Wilkins, Philadelphia

Yamamoto N, Watabe Y, Iwamoto M et al (2014) A case of mucoepidermoid carcinoma with clear cell components occurring in retromolar region. The Bulletin of Tokyo Dental College 55:25–31

Anatomy and Variations of the Mental Foramen

6

Joe Iwanaga and Paul J. Choi

6.1 Introduction

The mental foramen (MF) is the exit of the mandibular canal and a critical structure for implant surgery, periodontal surgery, periapical surgery, and local anesthesia. The mental neurovascular bundles emerge from the mental foramen, and dentists have to avoid injuring these anatomical structures. Therefore, determining the position of the mental foramen can help to preclude iatrogenic complications during the aforementioned procedures. Several textbooks and scientific papers have described the normal anatomy of the mental foramen; however, there are many variations that must be noted for better clinical practice so that unnecessary complications can be avoided. In this chapter, the normal anatomy and the variations of the mental foramen are reviewed, and their clinical relevance is discussed.

6.2 Mental Foramen/Nerve/Artery: Anterior Loop

Generally, the mental foramen (MF), which projects the mental neurovascular bundle, lies midway between the upper and lower borders of the body of the mandible bilaterally and is mostly located below the apex of the second premolar or between the first and second premolars. Studies of dry skulls indicate that the MF is positioned below the apex of the second premolar rather than between the two molars, but many of them fail to observe the distal curve of the root inside the bone. The MF is occasionally positioned between the second premolar and the first molar or below

J. Iwanaga (✉) · P. J. Choi
Seattle Science Foundation, Seattle, WA, USA
e-mail: joei@seattlesciencefoundation.org

J. Iwanaga, R. S. Tubbs (eds.), *Anatomical Variations in Clinical Dentistry*,
https://doi.org/10.1007/978-3-319-97961-8_6

the apex of the first premolar. Its distance from the alveolar ridge can change in association with bone loss caused by periodontitis and tooth loss, but the distance between the inferior border of the mandible and the inferior margin of the MF remains constant in the range 12–14 mm (von Arx et al. 2013; Al-Mahalawy et al. 2017). The bone loss causes an apparent rise in the position of the MF; therefore, in some patients, especially those who are edentulous, it is positioned on the upper border of the mandible. This variation perhaps accounts for the pain caused by denture compression. When mental nerve block is attempted in a patient whose mandible is edentulous with remarkable bone loss, the needle insertion site should be higher than for a patient with less bone loss (Fig. 6.1).

The mental nerve is the terminal branch of the inferior alveolar nerve, which arises from the mandibular nerve. It emerges from the MF of the mandible and divides into three branches: mental, inferior labial, and angular. Many researchers (Kamijo 1967; Alsaad et al. 2003) have described the distribution of the mental nerve. Kamijo (1967) described oral anatomy in detail, providing a useful resource for enhancing anatomical knowledge among dentists. His book describes the mental nerve as usually dividing into two branches, the mental and the inferior labial; the small part of the inferior labial branch running toward the angular regions often becomes the angular branch. The present authors agree that the angular branch is tiny and could be counted as one of the branches of the inferior labial. Hu et al. (2007) described four mental nerve branches: angular, lateral inferior labial, medial inferior labial, and mental (Fig. 6.2). These four branches were further classified into five different types on the basis of their branching patterns. Won et al. (2014) used Sihler's staining technique to visualize the distribution of the branches of the mental nerve. They established that the inferior labial branches, i.e., medial and lateral, rather than the others, innervated most of the inferior labial skin. In the authors' experience, the inferior labial branches are much thicker than the other two (Fig. 6.3). Alsaad et al. (2003) dissected the mental nerve using a surgical microscope and used three-dimensional reconstruction to describe its cutaneous branches and distribution patterns.

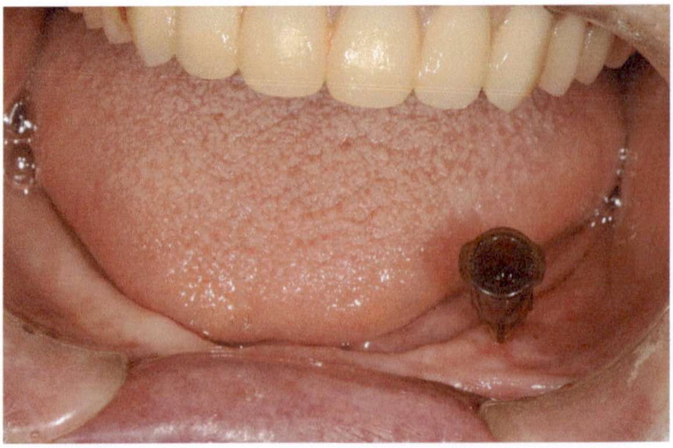

Fig. 6.1 Relatively high-positioned mental foramen in edentulous mandible

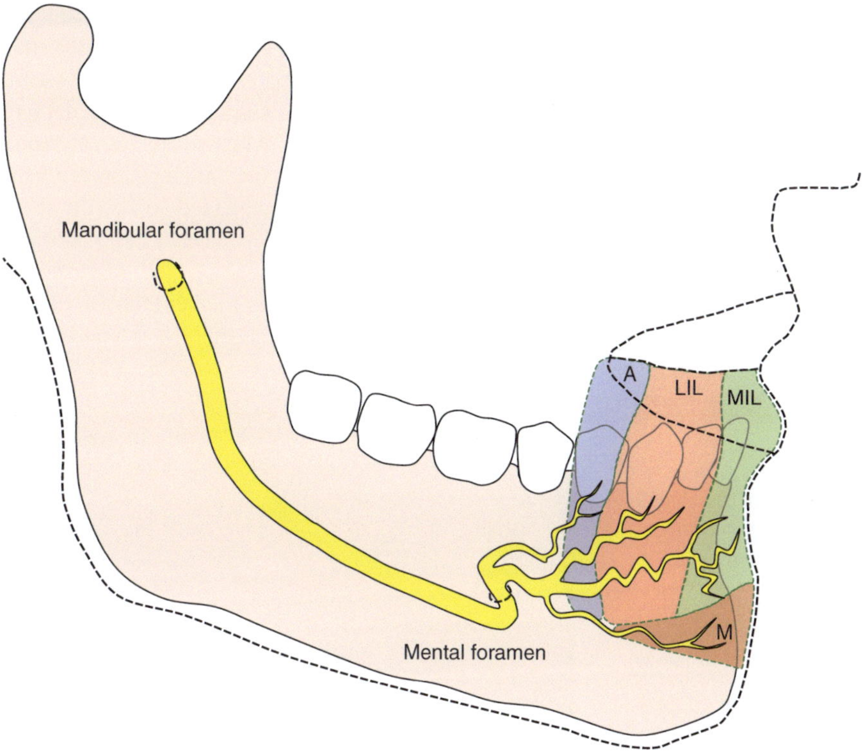

Fig. 6.2 Schematic drawing of four branches of the mental nerve. *A* the angular branch, *LIL* the lateral inferior labial branch, *MIL* the medial inferior labial branch, *M* the mental branch

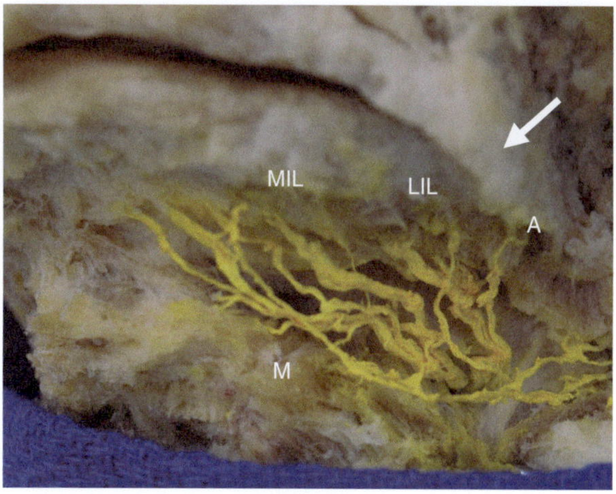

Fig. 6.3 Dissection of the left mental nerve. *A* the angular branch, *LIL* the lateral inferior labial branch, *MIL* the medial inferior labial branch, *M* the mental branch. Arrow: corner of the mouth

The depth of the running course of the mental nerve is clinically significant and can vary. After emerging from the MF, the inferior labial branch normally runs superficial to the incisivus labii inferioris and orbicularis oris muscles and toward the lower lip and lower labial glands. In some cases, the mental nerve runs deep to or pierces the incisivus labii inferioris (Fig. 6.4). Hence, when vestibular extension of the mandible is planned, the depth and width of the mucous incision are very important. Although no definitive landmark has been established so far, the main trunk of the inferior labial branch could run anteromedially beneath the mucosa of the lower lip.

The mental artery has long been depicted as similar to or thicker than the mental nerve. However, the mean external diameter of the mental artery is 0.5 ± 0.2 mm (range: 0.1–1.0 mm) (Iwanaga et al. 2016) (Fig. 6.5). Injury to this thin artery rarely

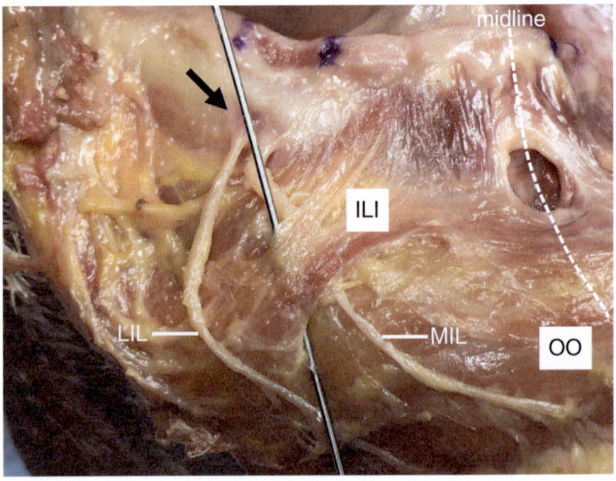

Fig. 6.4 The medial inferior labial branch of the mental nerve emerging from the mental foramen (arrow), piercing the incisivus labii inferioris muscle and running over the orbicularis oris muscle (right mandible). *ILI* incisivus labii inferioris muscle, *LIL* the lateral inferior labial branch, *MIL* the medial inferior labial branch, *OO* the orbicularis oris muscle

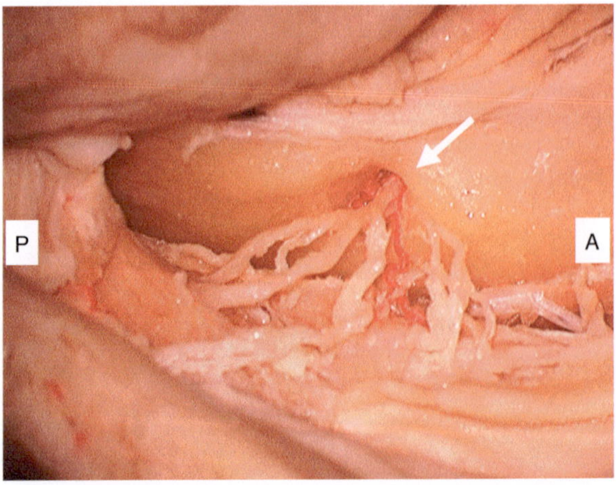

Fig. 6.5 Dissection of the mental nerve and mental artery emerging from mental foramen (arrow). Note that the mental artery is much thinner than the mental nerve

entails a huge hemorrhage, but it should be avoided. The mental artery supplies the mental and inferior labial regions and often anastomoses with the branches of the inferior labial and submental arteries.

The anterior loop (AL) has many variants and is one of the most significant structures to note before implants are placed in the premolar region of the mandible. It is present in 28–94% of the population. Temporary or permanent changes, or disturbance of sensation after intraforaminal implant placement, have been reported in 8.5–24% and 7%, respectively. Although most studies have reported only the length of the AL, Filo et al. (2014) measured the AL two-dimensionally, i.e., in the horizontal and vertical planes. Twenty-one percent (146/690) of patients had no anterior extension of the AL. Among the remaining 79%, approximately 96% had an anterior extension less than 3 mm long, while 4% had an extension up to 5.6 mm. The mean anterior extension of the AL was 1.2 mm. Filo et al. also measured a caudal extension of the AL (i.e., the distance between the MF and the inferior border of the mandibular canal). All patients had a caudal extension mean length of 4.1 mm (0.3–8.9 mm). Neiva et al. (2004) reported the anterior extension of the AL to be 11 mm at most. This entails a warning that the anterior loop should be considered even when implant surgery is planned in the intraforaminal area. Panoramic radiography can usually show the position of the MF; however, it often demonstrates the AL pattern inaccurately.

6.3 Accessory Mental Foramen/Nerve/Artery

The accessory mental foramen (AMF) is defined as any foramen in addition to the MF in the lateral body of mandible (Singh and Srivastav 2011). Some have described the AMF as a small foramen in the area surrounding the MF. Naitoh et al. (2011) and Sisman et al. (2014), using cone-beam computed tomographic (CBCT) images, described AMFs as foramina continuous with the mandibular canal and smaller than the MF on each side. According to Fuakami et al. (2011), a foramen associated with a branch of the mental nerve with no connection to the mandibular canal was observed in only a few CBCT images. These findings show that it is difficult to define the AMF using only radiological methods (Fuakami et al. 2011). Pancer et al. (2014) defined the AMF, in clinical terms, as a foramen connected to the mandibular canal on CBCT and continuous with the inferior alveolar nerve. Oliveira-Santos et al. (2011) and von Arx et al. (2014) described the difference between an AMF and a double MF. These two reports described a double MF as ≥50% of the size of the ipsilateral MF. According to Iwanaga et al. (2016), a double MF is similar in size and located adjacent to the MF (with a range of 1.5–4.5 mm), and it can contain one of the inferior labial branches (Fig. 6.6).

Recently, CBCT has provided clear images useful for diagnosis and planning of dental surgery. However, it is still not possible to identify the traveling course of the mental nerve (MN) and accessory MN (AMN) fully from radiographic images. A case was reported in which an AMF was detected on CBCT, but not under a gross anatomical observation; only an artery was grossly observable (Fuakami et al.

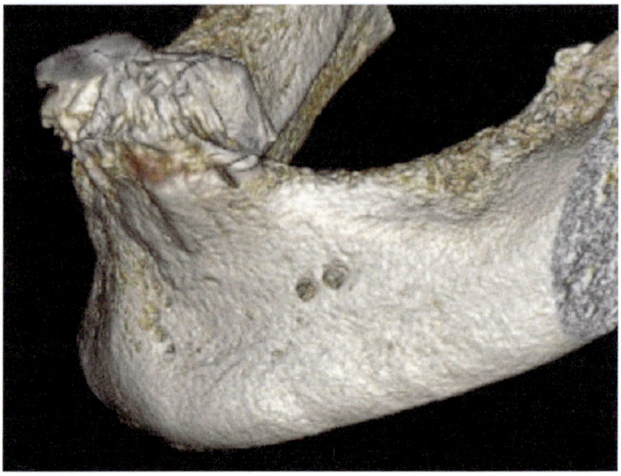

Fig. 6.6 Double mental foramen

2011). Although the definition proposed by Pancer et al. (2014) appears better from a clinical point of view, it needs to be confirmed by gross anatomical dissections or surgical procedures. Anatomical and clinical findings provide the only definitive confirmation of the presence of AMFs by demonstrating the neurovascular bundles as a whole when they exit the foramina.

The prevalence of AMFs ranges between 2.0% and 14.3% (Sisman et al. 2014; Iwanaga et al. 2016). There are no significant sex differences (Oliveira-Santos et al. 2011; Kalender et al. 2012; Sisman et al. 2014). AMFs have been observed by various methods. Gross anatomical observation of the dry skull, the classical and most reliable means, indicates a prevalence of 5.5–13% (Singh and Srivastav 2011; Udhaya et al. 2013). However, since we cannot see inside the mandible using this method, some of the grossly observed small foramina could mistakenly appear to connect with the mandibular canal. The number of CT studies has drastically increased during the past decade (Naitoh et al. 2009; Oliveira-Santos et al. 2011; Kalender et al. 2012; Orhan et al. 2013; Iwanaga et al. 2016), particularly with the development of CBCT, and the prevalence of AMFs measured by these methods is 2.0–14.3%. There seems to be no statistical difference between gross anatomical and CT studies in the prevalence of AMFs. According to Orhan et al. (2013), the prevalence of AMFs in children using CBCT is 6.3%. Panoramic radiography was also compared with CBCT (Naitoh et al. 2011; Neves et al. 2014). According to Naitoh et al. (2011), 48.6% of AMFs or bifurcations thereof could be visualized on panoramic radiography, while Neves et al. (2014) reported them in 7.4% of CBCT images and only 1.2% of panoramic radiographs and inferred a significant difference in prevalence between the two imaging modalities. In the authors' experience, panoramic radiography has extremely low reliability for detecting the AMF. We should consider that AMFs generally cannot be detected in panoramic radiographs unless they are large enough. Anatomical observation of dry skulls alone could result in variable outcomes since nutrient foramina could be counted as AMFs. Periapical fistulae or bony defects could also be mistaken for AMFs. Only panoramic observation results in a significantly lower rate of detection of AMFs according to the reviewed literature.

All previous studies have reported that AMFs are more often observed unilaterally than bilaterally (Oliveira-Santos et al. 2011; Sisman et al. 2014; Kalender et al. 2012), except for Iwanaga et al. (2016). Bilateral AMFs are relatively rare and have been described in 6–56% of all AMF cases. Unilateral AMFs occur in single, double, or triple forms. Mandibles with triple AMFs are very rare.

The accessory mental artery has rarely been described. However, Iwanaga et al. (2016) investigated it and reported that its mean external diameter was 0.6 mm ± 0.1 mm (range: 0.5–0.8 mm). In one of their cases, the AMF contained only an artery and no nerves.

6.3.1 Location of the MF and AMFs

The reported locations of the MF and AMFs differ among published articles. Some studies report that AMFs are frequently located posteroinferior to the MF (Sisman et al. 2014; Orhan et al. 2013; Neves et al. 2014), while others report them as anteroinferior (Kalender et al. 2012) or posterior to it. According to Iwanaga et al. (2015), a site posterior to the MF was more common than its anterior counterpart overall. However, one study showed the positional mapping of 20 AMFs in relation to their sizes (Iwanaga et al. 2016). Smaller AMFs tended to be located posterosuperior to and far from the MF, while larger ones were more often located superior or anterior and close to the MF. This means that if the gingiva and periosteum are elevated coronally to apically in a surgical procedure without the possible presence of AMFs being considered, the large foramen could be manipulated and its contents injured even before the MF is detected (Fig. 6.7). Therefore, CBCT is often not recommended for planning not only implant surgery but also routine periodontal surgery.

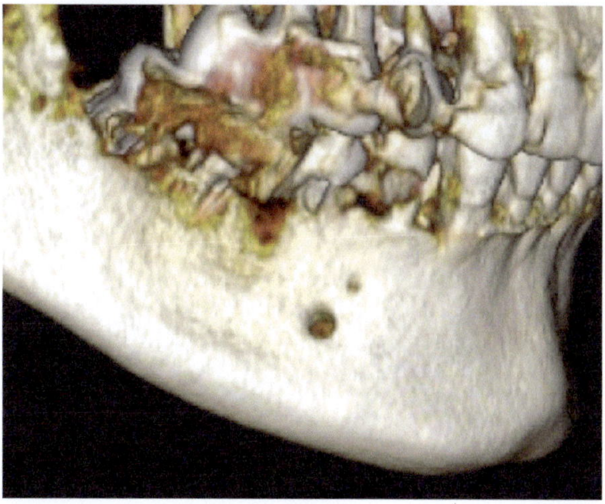

Fig. 6.7 Accessory mental foramen located superior to the mental foramen

6.3.2 Distance Between the MF and AMFs

The distance between the MF and AMFs ranges from 0.7 mm to 8.7 mm (Naitoh et al. 2009; Kalender et al. 2012; Sisman et al. 2014; Iwanaga et al. 2016).

6.3.3 Size of the MF and AMFs

Many studies have measured the vertical and horizontal diameters of AMFs. However, only a few have measured their areas. According to Naitoh et al. (2009), the mean area of 15 AMFs was 1.7 mm^2 and that of the ipsilateral MFs was 7.5 mm^2, whereas the mean area of MFs without ipsilateral AMFs was 9.4 mm^2. The presence of AMFs did not influence MF size significantly. MNs, regardless of whether ipsilateral AMNs are present, have almost the same number of nerve fibers before they branch off from the mandible. In addition, since the thickness of the nerve bundle in the AMN can affect the sizes of the MF and AMF, the larger the AMF, the smaller the ipsilateral MF.

When there is a large AMF similar in size to the MF, the term "double mental foramen" is used. The AMF can contain a thick nerve bundle, i.e., an inferior labial branch. One should avoid injuring double MFs. The MF ipsilateral to a large AMF tends to be much smaller than the contralateral one even on panoramic radiographs. When there is an obvious size difference between the right and left MFs, one should suspect the existence of an AMF (Fig. 6.8).

According to Toh et al. (1992) and Iwanaga et al. (2016), the nerve bundles arising from the AMF are part of the MN and branch off before the main trunk emerges from the MF. Therefore, there appear to be fewer nerve fibers in an MN with an ipsilateral AMN than in one without and in an MF with an ipsilateral AMF than one without. Iwanaga et al. (2017) reported that AMFs smaller than 1.3 mm^2 were not clearly identifiable on surface-rendered images, though they could be confirmed in cross-sectional images (Fig. 6.9).

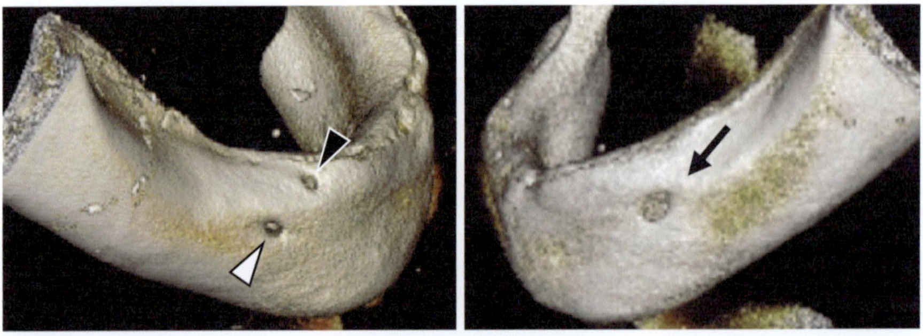

Fig. 6.8 Difference in size between mental foramen (white arrowhead) and contralateral mental foramen (arrow) when a large accessory mental foramen (black arrowhead) is present in the same specimen. Note this accessory mental foramen is almost the same size as the mental foramen (it could be called a "double mental foramen")

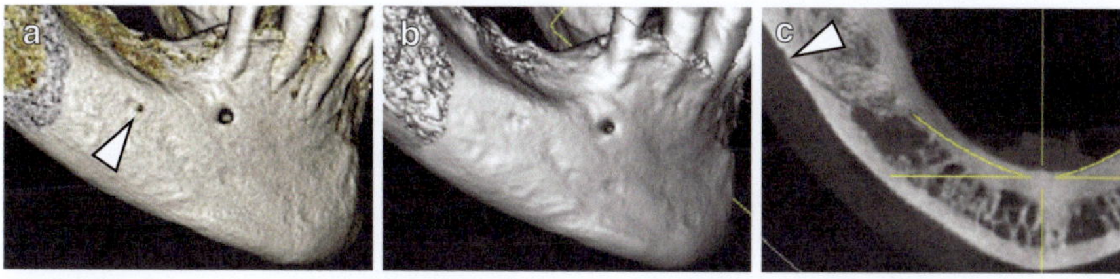

Fig. 6.9 Small accessory mental foramen (arrowheads) detected on reconstructed images using a DICOM viewer (**a**) and cross-sectional images (**c**), but not in surface-rendered images (**b**)

6.3.4 Distribution of the MN and AMNs

The different distributions of the accessory mental nerve are related to the positions of the AMF and MF, since the fibers of the AMN supplement the areas to which the MN fibers do not reach. Clinically, an AMN can cause technical difficulties in some procedures such as local anesthesia, implant surgery, periapical surgery, trauma surgery, and neurectomy. Concepcion and Rankow (2000) described a case in which the AMF was located posterosuperior to the MF, i.e., close to the operative field of a periapical surgery. The patient suffered a mild tingling sensation around the corner of the lower lip, which persisted for 3 weeks after the surgery. Kulkarni et al. (2011) reported a case with an AMN in the buccal aspect of the alveolar bone observed during an implant placement. Careful dissection left the AMN uninjured in this case. von Arx et al. (2014) presented clinical cases of periapical surgery during which an AMF and an AMN were noted. In all four cases, the soft tissue bundles arising from the AMF were excised and showed similar histological findings. Three of those patients had no postoperative complication such as paresthesia. However, paralysis of the perioral soft tissues was reported in one case in which the AMF was large, i.e., over half the size of the MF, and positioned posterosuperiorly to the MF. Pancer et al. (2014) located an AMF during implant surgery, which led them to alter the osteotomy site to preclude complications that could have arisen from injuring the AMF. After the surgery, the patient experienced no signs of paresthesia or atypical symptoms. According to López and Diago (2006), AMFs are among the anatomical variations that can cause failure of dental anesthesia, though the current literature presents no evidence supporting this claim. Ramadhan et al. (2010) detected AMFs during a trauma surgery. When they attempted to reposition and fixate the mandible via an intraoral approach, they found three different foramina around the MF giving rise to three nerve bundles. They therefore decided to use a submandibular approach, which had an uneventful outcome. Two cases of neurectomy to manage trigeminal neuralgia have been reported in patients with AMNs. Both the AMN and the MN were excised, after which the patients were completely free of neuropathic pain. Gamoh et al. (2014) reported a case in which AMFs were initially misdiagnosed as tumors in a private clinic. In this case, even the most experienced oral and maxillofacial radiologist could not spot the AMFs using panoramic radiography.

6.4 Absence of Mental Foramen

Bilateral (Lauhr et al. 2015; Hasan 2013; Matsumoto et al. 2013) and unilateral (Ulu et al. 2016; da Silva Ramos et al. 2011) absence of the MF is rare but has been identified by CBCT, dry mandible observation, or surgical exploration in vivo. However, no sensory disturbance of the mental region has been reported in these subjects (da Silva Ramos et al. 2011; Ulu et al. 2016). Hasan (2013) reviewed the literature describing the incidence of absent MF, showing that 0.08% (1/1250) to 0.95% (1/105) of mandibles lacked a unilateral MF. The literature review revealed no reported cases of sensory disturbance. Even when no MF can be found by CBCT, the mental region can receive compensatory sensory input from other nerve fibers. Therefore, mental nerve block for patients whose MF is absent should be replaced by infiltration anesthesia and inferior alveolar nerve block.

6.5 Other Clinical Considerations

From the clinical perspective, the MF is a particularly significant anatomical structure since it frequently appears in a surgical field. Hence, dentists should strive to avoid manipulating this small structure to prevent nerve paralyses. If the full thickness of the periosteum is elevated, a nerve injury ensues even when the MF is located since the mental neurovascular bundle is covered by the periosteum and epineurium. This also indicates that the mental nerve cannot be injured as long as the periosteum is intact (Fig. 6.10). Some dentists perform surgery with minimal elevation of the periosteum and exposure of the MF in order to avoid injuring the mental neurovascular bundles. However, this occasionally results in excessive retraction of the flap, which could injure both the periosteum and the neurovascular bundles. It is crucial to secure enough of the operative field for the periosteum to be manipulated gently. A very thin AMN resembling a "fine thread" can be cut (Fig. 6.11), but a thick AMN should not be cut unless the procedure is for resection of tumors or other serious diseases. When we observe one side of the MF in panoramic radiography, we should also pay attention to the contralateral MF. If there is a clear size difference between the right and left MFs, this could indicate that an AMF is present.

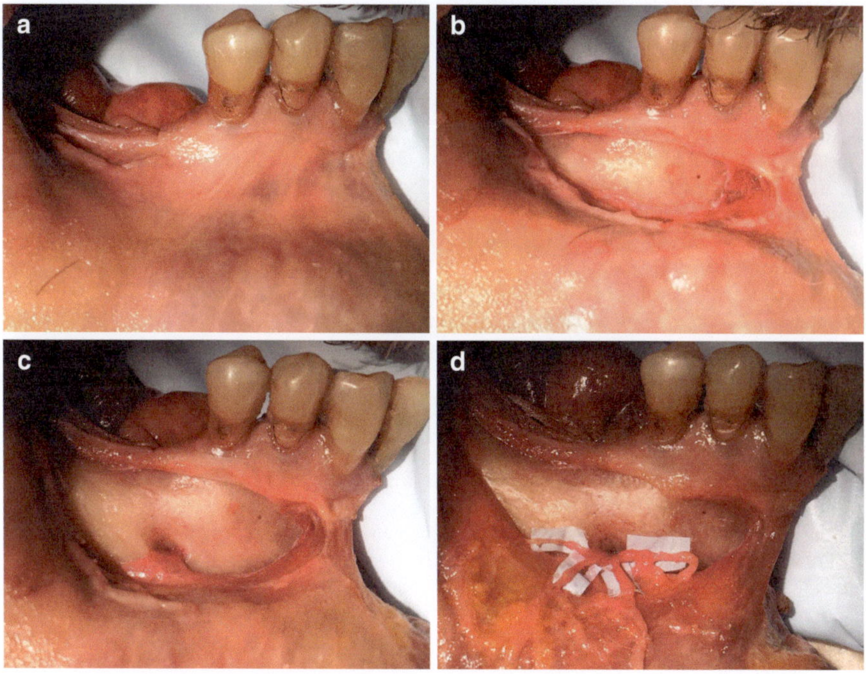

Fig. 6.10 Relationship between the mental nerve and the periosteum. (**a**) Before mucosa incision. (**b**) Elevating the periosteum but not yet identifying the mental foramen. (**c**) Elevating the periosteum; mental foramen identified. Note that the mental foramen is still covered with periosteum. (**d**) The periosteum and epineurium are removed and the mental nerve exposed

Fig. 6.11 Very thin accessory mental nerve (arrowheads) resembling a "fine thread"

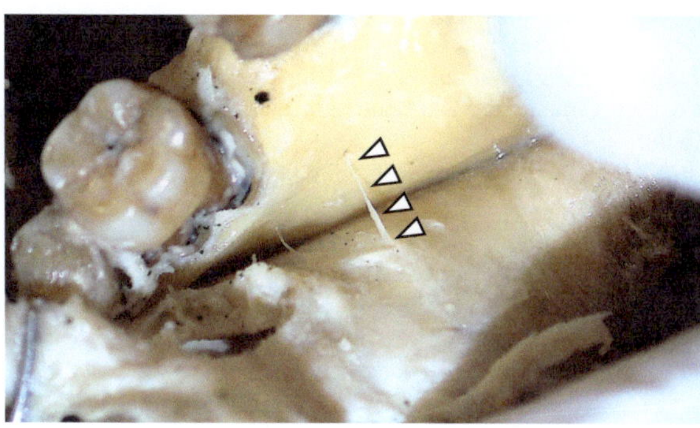

References

Al-Mahalawy H, Al-Aithan H, Al-Kari B et al (2017) Determination of the position of mental foramen and frequency of anterior loop in Saudi population. A retrospective CBCT study. Saudi Dent J 29:29–35. https://doi.org/10.1016/j.sdentj.2017.01.001

Alsaad K, Lee TC, McCartan B (2003) An anatomical study of the cutaneous branches of the mental nerve. Int J Oral Maxillofac Surg 32:325–333

Concepcion M, Rankow HJ (2000) Accessory branch of the mental nerve. J Endod 26:619–620

da Silva Ramos LMP, Capelozza ALÁ, Rubira-Bullen IRF (2011) Absence and hypoplasia of the mental foramen detected in CBCT images: a case report. Surg Radiol Anat 33:731–734

Filo K, Schneider T, Locher MC et al (2014) The inferior alveolar nerve's loop at the mental foramen and its implications for surgery. J Am Dent Assoc 145:260–269. https://doi.org/10.14219/jada.2013.34

Fuakami K, Shiozaki K, Mishima A et al (2011) Detection of buccal perimandibular neurovascularisation associated with accessory foramina using limited cone-beam computed tomography and gross anatomy. Surg Radiol Anat 33:141–146

Gamoh S, Mori Y, Nakatsuka M, Akiyama H, Ogawa Y, Iwai Y, Morita S, Shimizutani K (2014) Accessory Mental Foramen Misdiagnosed as Radiolucent Tumour by Conventional Dental Radiography. Open J Radiol 4:173–176

Hasan T (2013) Bilateral caroticoclinoid and absent mental foramen: rare variations of cranial base and lower jaw. Ital J Anat Embryol 118:288–297

Hu KS, Yun HS, Hur MS et al (2007) Branching patterns and intraosseous course of the mental nerve. J Oral Maxillofac Surg 65:2288–2294

Iwanaga J, Saga T, Tabira Y et al (2015) The clinical anatomy of accessory mental nerves and foramina. Clin Anat. https://doi.org/10.1002/ca.22597

Iwanaga J, Watanabe K, Saga T et al (2016) Accessory mental foramina and nerves: application to periodontal, periapical, and implant surgery. Clin Anat 4(29):493–501. https://doi.org/10.1002/ca.22635

Iwanaga J, Watanabe K, Saga T et al (2017) Undetected small accessory mental foramina using cone-beam computed tomography. Cureus 9(5):e1210

Kalender A, Orhan K, Aksoy U (2012) Evaluation of the mental foramen and accessory mental foramen in Turkish patients using cone-beam computed tomography images reconstructed from a volumetric rendering program. Clin Anat 25:584–592

Kamijo Y (1967) Oral anatomy, 1st edn. Anatom Co., Tokyo

Lauhr G, Coutant JC, Normand E et al (2015) Bilateral absence of mental foramen in a living human subject. Surg Radiol Anat 37:403–405. https://doi.org/10.1007/s00276-014-1347-x

López AB, Diago MP (2006) Failure of locoregional anesthesia in dental practice. Review of the literature. Med Oral Patol Oral Cir Bucal, 11

Matsumoto K, Araki M, Honda K (2013) Bilateral absence of the mental foramen detected by cone-beam computed tomography. Oral Radiol 29:198–201

Naitoh M, Hiraiwa Y, Aimiya H et al (2009) Accessory mental foramen assessment using cone-beam computed tomography. Oral Surg Oral Med Oral Pathol Oral Radiol Endod 107:289–294

Naitoh M, Yoshida K, Nakahara K et al (2011) Demonstration of the accessory mental foramen using rotational panoramic radiography compared with cone-beam computed tomography. Clin Oral Implants Res 22:1415–1419

Neiva RF, Gapski R, Wang H-L (2004) Morphometric analysis of implant-related anatomy in Caucasian skulls. J Periodontol 75:1061–1067

Neves FS, Nascimento MCC, Oliveira ML et al (2014) Comparative analysis of mandibular anatomical variations between panoramic radiography and cone beam computed tomography. Oral Maxillofac Surg 18:419–424

Oliveira-Santos C, Souza PH, De AzambujaBerti-Couto S et al (2011) Characterisation of additional mental foramina through cone beam computed tomography. J Oral Rehabil 38:595–600

Orhan AI, Orhan K, Aksoy S et al (2013) Evaluation of perimandibular neurovascularization with accessory mental foramina using cone-beam computed tomography in children. J Craniofac Surg 24:e365–e369

Pancer B, Garaicoa-Pazmiño C, Bashutski JD (2014) Accessory mandibular foramen during dental implant placement: case report and review of literature. Implant Dent 23:116–124

Ramadhan A, Messo E, Hirsch JM (2010) Anatomical variation of mental foramen. Stomatologija 12:93–96

Singh R, Srivastav AK (2011) Evaluation of position, shape, size and incidence of mental foramen and accessory mental foramen in Indian adult human skulls. Int J Exp Clin Anat 5:23–29

Sisman Y, Sahman H, Sekerci AE et al (2014) Detection and characterization of the mandibular accessory buccal foramen using CT. Dentomaxillofac Rad 41:558–563

Kulkarni S, Kumar S, Kamath S et al (2011) Accidental identification of accessory mental nerve and foramen during implant surgery. J Indian Soc Periodontol 15:70

Toh H, Kodama J, Yanagisako M et al (1992) Anatomical study of the accessory mental foramen and the distribution of its nerve. Okajimas Folia Anat Jpn 69:85–88

Udhaya K, Saraladevi KV, Sridhar J (2013) The morphometric analysis of the mental foramen in adult dry human mandibles: A study on the South Indian population. J Clin Diagn Res 7:1547–1551

Ulu M, TarimErtas E, Gunhan F et al (2016) Unilateral absence of mental foramen with surgical exploration in a living human subject. Case Rep Dent 2016:1971925. https://doi.org/10.1155/2016/1971925

von Arx T, Friedli M, Sendi P et al (2013) Location and dimensions of the mental foramen: a radiographic analysis by using cone-beam computed tomography. J Endod 39:1522–1528. https://doi.org/10.1016/j.joen.2013.07.033

von Arx T, Lozanoff S, Bosshardt D (2014) Accessory mental foramina: anatomy and histology of neurovascularisation in four cases with apical surgery. Oral Surg 7:216–227

Won SY, Yang HM, Woo HS et al (2014) Neuroanastomosis and the innervation territory of the mental nerve. Clin Anat 27:598–602

7 Variant Anatomy of the Torus Mandibularis

Soichiro Ibaragi

7.1 Introduction

In Latin, "tori" means "to stand out" or "lump" (Reino et al. 1990). A torus mandibularis (TM) is a slow-growing bony protuberance located at the lingual side of the mandible in the premolar region above the mylohyoid line (Fig. 7.1). It consists of dense cortical bone with a small amount of bone marrow and is covered with thin mucosa and periosteum. Patients therefore occasionally suffer scalds or injuries to the mucosa of the TM by hot water or toothbrushing. A TM is usually found incidentally during a dental checkup as it usually exhibits no symptoms.

7.2 Etiology

The prevalence of TM is 6–58% in different populations (García-García et al. 2010; Morita et al. 2017). The etiology is not definitive but could be multifactorial involving environmental factors such as individual diet; occlusal stress, e.g., bruxism; and genetic factors; this is the most widely accepted hypothesis. However, the contribution of occlusal stress and bruxism to TM is still controversial; only a few studies have assessed oral and occlusal status (Suzuki and Sakai 1960; Eggen and Natvig 1986; Kerdpon and Sirirungrojying 1999; Loukas et al. 2013; Morita et al. 2017). Approximately 30% of TMs are estimated to be of genetic origin, while 70% are caused by environmental factors, mainly related to occlusal stress (Eggen 1989; García-García et al. 2010; Cortes et al. 2014; Auškalnis et al. 2015).

Radiographic examination shows that a TM is a radio-opaque mass continuous with the surrounding bone (Fig. 7.2). However, radiographic examination is not

S. Ibaragi (✉)
Department of Oral and Maxillofacial Surgery, Okayama University Graduate School of Medicine, Dentistry and Pharmaceutical Sciences, Okayama, Japan
e-mail: sibaragi@md.okayama-u.ac.jp

J. Iwanaga, R. S. Tubbs (eds.), *Anatomical Variations in Clinical Dentistry*,
https://doi.org/10.1007/978-3-319-97961-8_7

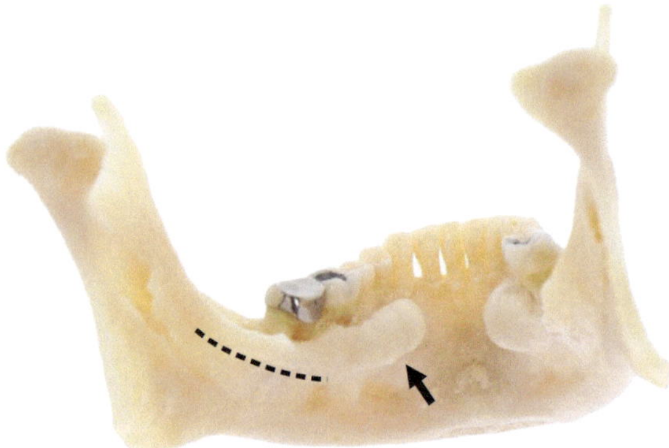

Fig. 7.1 Torus mandibularis (arrow) and mylohyoid line (dotted line)

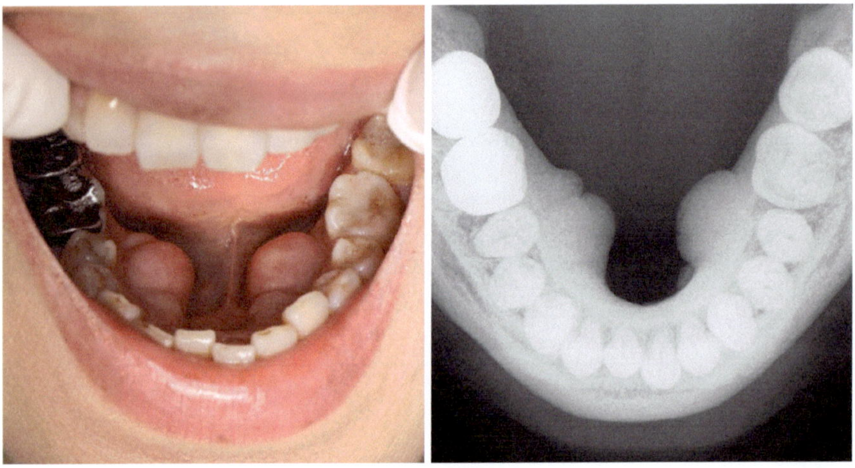

Fig. 7.2 Oral photograph and occlusal radiograph of torus mandibularis

very useful for diagnosis because TMs are generally diagnosed simply by clinical examination. Computed tomography is a better choice for estimating the bone height and thickness before surgery (Fig. 7.3). Histopathological examination shows TMs to have a structure similar to normal bone.

7.3 Morphology

TMs are classified in terms of the side and number of lesions (Fig. 7.4): (1) unilateral solitary, (2) bilateral solitary, (3) unilateral multiple, (4) bilateral multiple, and (5) bilateral combined. Most TMs are bilateral solitary (39.76%). The average size of a TM is 10.0–10.9 × 6.49–9.0 mm (García-García et al. 2010). Most are located on the lingual side of the mandible bilaterally, above the mylohyoid line (Scrieciu et al. 2016). A large TM can cause difficulty in swallowing and tongue movement (Fig. 7.5).

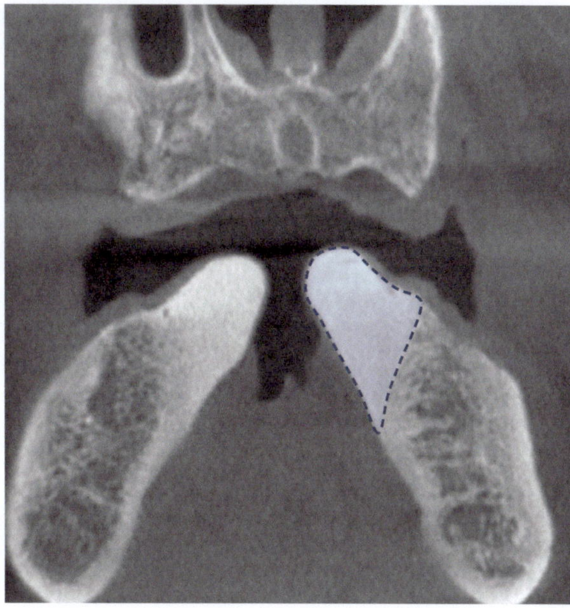

Fig. 7.3 A coronal CBCT image of tori mandibularis

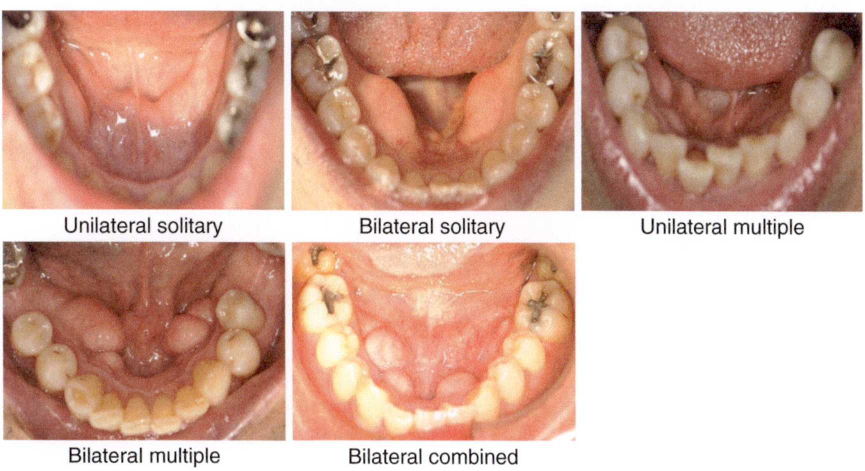

Fig. 7.4 Classification of torus mandibularis

7.4 Sex, Age, Ethnic Groups

Most studies show no statistically significant gender differences. A TM rarely appears before the first decade of life; it usually becomes prominent in the second or third decade. The average age of patients with TM is 34–39.2 years

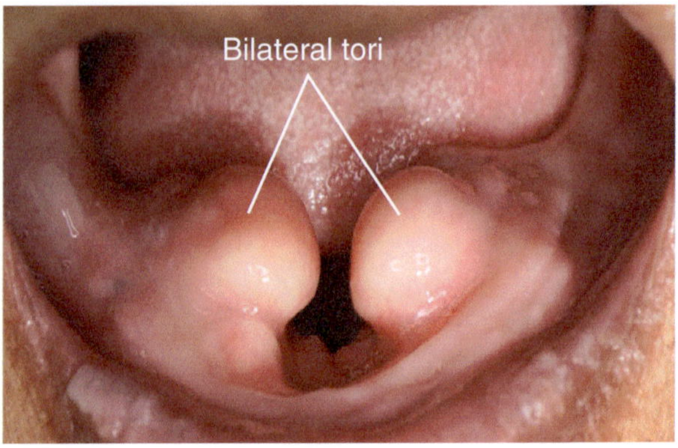

Fig. 7.5 Large bilateral torus mandibularis

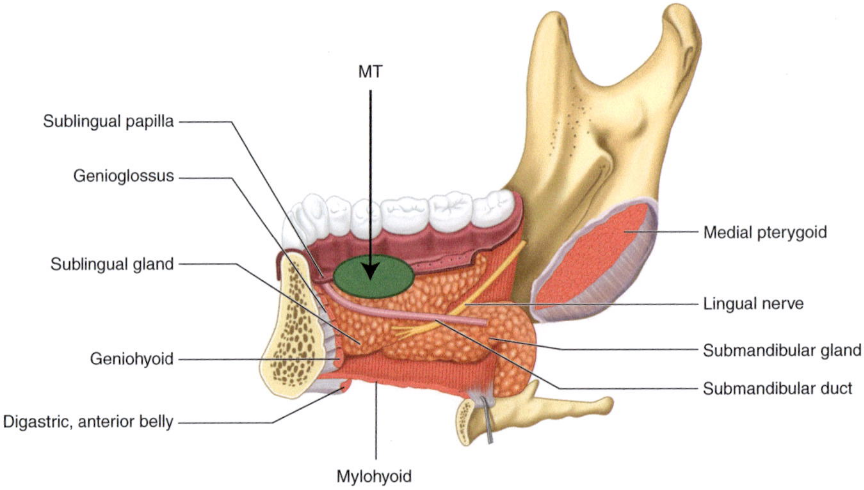

Fig. 7.6 Schematic drawing of torus mandibularis and adjacent structures

(García-García et al. 2010). TM appears commonly in certain ethnic groups and countries including Eskimos, Japanese, and Mongols and in the United States (García-García et al. 2010).

7.5 Anatomy of Adjacent Organs (Fig. 7.6)

7.5.1 Salivary Duct

The excretory duct of the submandibular gland, called Wharton's duct, opens at the sublingual caruncle. The sublingual gland has small Rivinus's ducts and may have a larger Bartholin's duct, which open at the sublingual fold and sublingual caruncle, respectively.

7.5.2 Sublingual Artery, Submental Artery, and Branches

The sublingual artery arises from the lingual artery. It runs upward and immediately curves downward, crossing the hypoglossal nerve inferiorly, and goes forward beneath the hyoglossus and upward to the tongue. The submental artery branches from the facial artery inferior and medial to the inferior border of the mandible. One or both of these arteries gives rise to the vessel entering the lingual foramina (He et al. 2017).

7.5.3 Lingual Nerve

The lingual nerve runs forward between the mandibular ramus and medial pterygoid muscle, reaches the border of the tongue beneath superior pharyngeal constrictor muscle, and then runs forward beside the submandibular gland on the mylohyoid. Then it crosses Wharton's duct inferiorly, laterally to medially, and goes along the tongue margin to the tongue tip.

7.6 Surgical Technique

Removal of a TM is not always indicated. It is indicated in cases of significant growth that could disturb mastication and speech or constrict a prosthesis.

Other reasons reported are traumatic ulcers, food retention, cancerophobia, and source of autogenous bone for graft (Hassan et al. 2015).

In most cases, local anesthesia is sufficient. The anesthetic is administrated by infiltration and/or the inferior alveolar nerve block. The incision is made on the mandibular ridge or at the necks of the teeth above the TM (Figs. 7.7 and 7.8). A mucoperiosteal flap is elevated by periosteotomes until the TM is exposed. Note that the mucosa and periosteum covering the TM are very thin and easily torn. The excision is performed using a fissure bur or a osteotome. Afterward, a suture is made

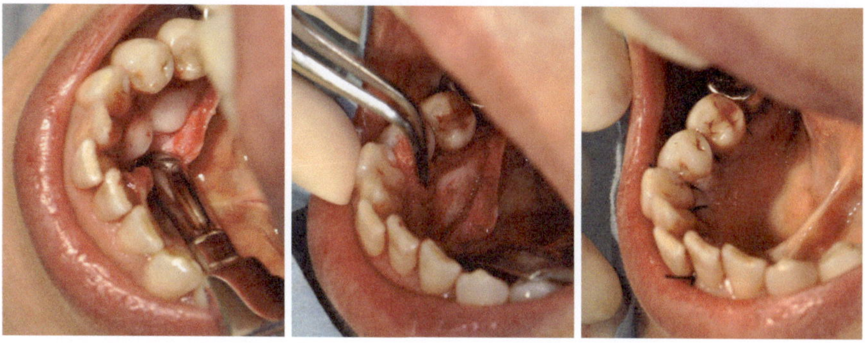

Fig. 7.7 Intraoperative photographs of removal of torus mandibularis

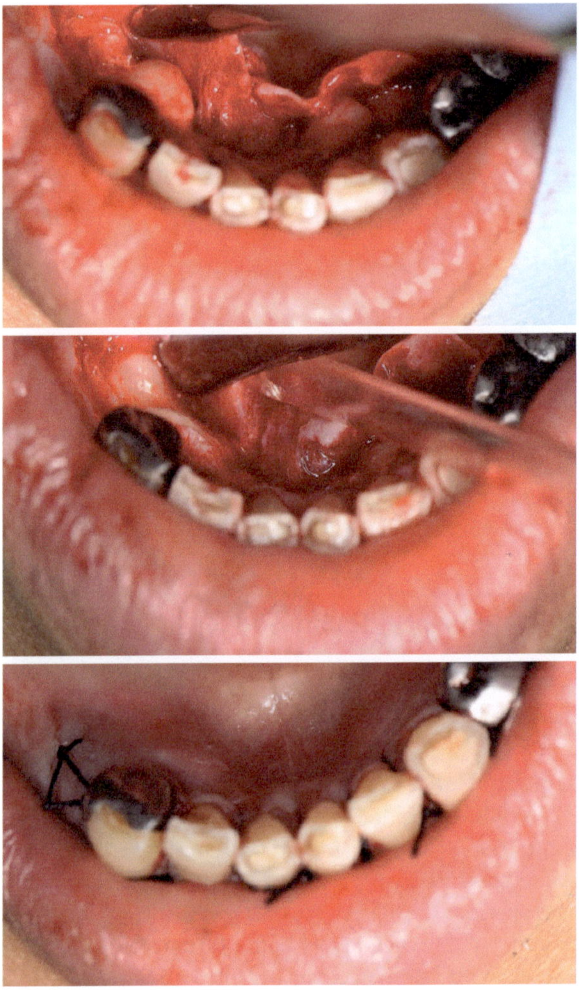

Fig. 7.8 Intraoperative photographs of removal of torus mandibularis

and the wound can be protected from mechanical force by surgical cement or adjustment of a prosthesis.

7.7 Complications

Possible surgical complications, though uncommon, include fracture of the mandible, direct injection of anesthetic into a blood vessel, swallowing or inhalation of the bone fragments, devitalization of neighboring teeth, injury to the roots of adjacent teeth, submandibular duct injuries, lingual nerve injury, mucosal laceration, and poor adaptation of the flap (Reino et al. 1990; Martínez-González 1998; Donado 1998; Barker et al. 2001; García-García et al. 2010).

References

Auškalnis A, Rutkūnas V, Bernhardt O et al (2015) Multifactorial etiology of Torus mandibularis: study of twins. Stomatologija 17:35–40

Barker D, Walls AWG, Meechan JG (2001) Case report: ridge augmentation using mandibular tori. Br Dental J 190:474

Cortes ARG, Jin Z, Morrison MD et al (2014) Mandibular tori are associated with mechanical stress and mandibular shape. J Oral Maxillofac Surg 72:2115–2125

Donado M (1998) Pre-prosthetic Surgery. In: Donado M (ed) Cirugíabucal Patología y técnica. [Oral Surgery. Pathology and Technique], 2nd edn. Masson, Barcelona, pp 481–510

Eggen S (1989) Torus mandibularis: an estimation of the degree of genetic determination. Actaodontologicascandinavica 47:409–415

Eggen S, Natvig B (1986) Relationship between torus mandibularis and number of present teeth. Scand J Dent Res 94:233–240

García-García AS, Martínez-González JM, Gómez-Font R et al (2010) Current status of the torus palatinus and torus mandibularis. Med Oral Patol Oral Cir Bucal 15:e353–e360

Hassan KS, Al-Agal A, Abdel-Hady AI et al (2015) Mandibular tori as bone grafts: an alternative treatment for periodontal osseous defects-clinical, radiographic and histologic morphology evaluation. J Contemp Dent Pract 16:192–200

He P, Truong MK, Adeeb N et al (2017) Clinical anatomy and surgical significance of the lingual foramina and their canals. Clin Anat 30:194–204

Kerdpon D, Sirirungrojying S (1999) A clinical study of oral tori in southern Thailand, prevalence and the relation to parafunctional activity. Eur J Oral Sci 107:9–13

Loukas M, Hulsberg P, Tubbs RS et al (2013) The tori of the mouth and ear: a review. Clin Anat 26:953–960

Martínez-González JM (1998) Tumoresbenignos de los maxilares [Benigntumors of the maxilla]. In: Donado M (ed) CirugíaBucal. PatologíayTécnica, 2th edn. Masson, Barcelona, pp 627–639

Morita K, Tsuka H, Shintani T et al (2017) Prevalence of torus mandibularis in young healthy dentate adults. J Oral Maxillofac Surg 75:2593–2598

Reino OC, Galera JP, Martin JPC (1990) Surgical techniques for the exeresis of torus, both palatal and mandibular. Rev Actual Odontoestomatol Esp 50:47–50

Scrieciu M, Mercuţ VERONICA, Mercuţ RĂZVAN et al (2016) Morphological and clinical characteristics of the torus palatinus and torus mandibularis in a sample of young and adults' Romanian people. Rom J Morphol Embryol 57:139–144

Suzuki M, Sakai T (1960) A familial study of torus palatinus and torus mandibularis. Am J Phys Anthropol 18:263–272

Part III

Maxillary Sinus

Anatomy and Variations of the Floor of the Maxillary Sinus

8

Katsuichiro Maruo, Charlotte Wilson, and Joe Iwanaga

8.1 Introduction

The maxillary sinus is one of the paranasal sinuses along with the frontal, ethmoidal, and sphenoidal sinuses. It is located above the apical roots of the maxillary molars and is an important structure in fields such as endodontics, implant dentistry, and oral surgery. In planning for dental implant treatment, the distance between the alveolar ridge and the sinus floor, the existence of pathological features, and the septum in the maxillary sinus are of great importance. For example, a periapical lesion or penetration of the sinus floor membrane after tooth extraction can cause maxillary sinusitis. In this chapter, the anatomy, structure, and variations of the maxillary sinus are reviewed, and the sinus floor elevation procedure is introduced.

8.2 Development

The maxillary sinus develops embryologically at about the 10th week of gestation. The cavity expands by the fifth month and pneumatizes at birth. The infant's maxillary sinus is approximately 8 × 4 × 4 mm, but it expands rapidly during the first year of postnatal life. It grows throughout childhood until approximately 12–14 years of age. Expansion is particularly rapid during eruption of the primary and permanent teeth and the corresponding growth of the alveolar ridge

K. Maruo (✉)
Sangenjaya Maruo Dental Clinic, Setagaya-ku, Tokyo, Japan

C. Wilson
University of Colorado, Denver, CO, USA
e-mail: Charlotte.Wilson@colorado.edu

J. Iwanaga
Seattle Science Foundation, Seattle, WA, USA
e-mail: joei@seattlesciencefoundation.org

J. Iwanaga, R. S. Tubbs (eds.), *Anatomical Variations in Clinical Dentistry*,
https://doi.org/10.1007/978-3-319-97961-8_8

(Lorkiewicz-Muszyńska et al. 2015). Pneumatization of the maxillary sinus can continue after adolescence owing to the eruption of the third molars and extraction or loss of the maxillary molars (van den Bergh et al. 2000).

8.3 Physiology and Function

The maxillary sinus floor membrane is made of respiratory mucosa: a multilayered cylindrical epithelium comprising a surface layer of ciliated and unciliated cylindrical cells, basal cells, muciparous beaker cells, an underlying basal membrane, and thick lamina propria containing vascular and glandular layers and the periosteum. Beaker cells produce mucus that traps dust and particles, keeps the membrane moist, and helps humidify inhaled air. The ciliated columnar epithelium transports the secretions produced in the maxillary sinus to the ostium (Fig. 8.1). The lining of healthy sinuses is 0.2–0.8 mm thick with a relatively thin basement membrane. However, according to Eggesbø (2006), thickening of the membrane is common in asymptomatic patients so that the mucosal lining is normal when less than 4 mm. A bilayered secretory blanket coats the lining. The inner layer, called the sol, is thin, serous, and rich in proteins, immunoglobulins, and complement. The surface layer, called the gel, is viscous mucus that floats on top of the thinner sol layer. The cilia of the cells lining the sinus reach up through the sol layer and sweep the gel along so that any surface materials are driven toward the sinus ostium at 3–25 mm per minute. This transportation capacity is limited to secretions and extremely small foreign-body particles such as dust. The cilia cannot remove larger particles such as root residues. The mucus flows in a predictable stellate pattern, beginning from the inferior and lateral portion of the sinus floor and ending at the superior and medial

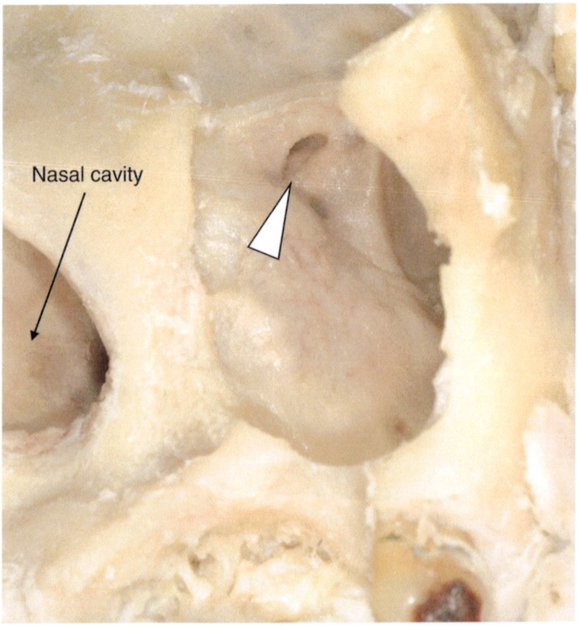

Fig. 8.1 Anteroinferior view of the ostium (arrowhead) on left maxillary sinus. Anterior wall of the maxillary sinus removed

sinus opening. The environment in the maxillary sinus is nearly sterile owing to the continuous beating pattern of the cilia (Bailey et al. 2006).

8.4 Nerves and Arteries of the Maxillary Sinus

Sensory innervation to the maxillary sinus is supplied by the posterior superior alveolar branch of V_2, and middle and anterior superior alveolar branches of the infraorbital nerve. Sympathetic innervation is via branches of the nasopalatine nerve and parasympathetic innervation via branches of the pterygopalatine ganglion. The maxillary sinus derives its blood supply predominately from the external carotid circulation via branches of the maxillary artery, mainly the infraorbital and the posterior superior alveolar arteries, but also from branches of the posterior lateral nasal and sphenopalatine arteries, which supply the middle portion of the sinus membrane. It is important to note that an extraosseous anastomosis sometimes occurs between the posterior superior alveolar artery and a terminal branch of the infraorbital artery. This anastomosis courses at a height of 23–26 mm from the alveolar margin and can cause hemorrhage when a flap is elevated during sinus graft surgery. On the other hand, there might be an intraosseous anastomosis, coursing at a distance of 19–20 mm from the alveolar margin. This anastomosis could not be avoided if it were within the lateral window outline; however, it is not significant and should not be treated with an electrosurgical unit but either ignored or managed by applying a hemostatic agent with light pressure. The venous drainage from the maxillary sinus is by the facial vein, the sphenopalatine vein, and the pterygoid venous plexus (Al-Faraje et al. 2013).

8.5 Function of Maxillary Sinus

The biological roles of paranasal sinuses are debated. However, a number of possible functions have been suggested, such as lightening the weight of the head, humidifying and heating inhaled air, increasing the resonance of speech, providing a buffer against facial trauma, and aiding the immune defense of the nasal cavity (Keir 2009).

8.6 The Relationships Between the Roots and Maxillary Sinus

During sinus pneumatization, the sinus floor moves inferiorly and becomes closer to the roots of the maxillary teeth (Fig. 8.2). In more than 50% of the population, the inferior sinus wall lies among the roots of the posterior maxillary teeth, creating a "hillock" (elevation of the sinus floor) or protrusion of the roots into the sinus (Hauman et al. 2002). In this case, the bone of the sinus floor is significantly less thick. However, most roots that protrude into the sinus are enveloped by a thin cortical layer on histological sections, and the rates of true perforation are only 14–28% (Wehrbein and Diedrich 1992). Many studies concern the vertical and horizontal relationships from the maxillary posterior root apices to the inferior wall of the sinus and their effect on endodontic treatment, implant placement, tooth extraction,

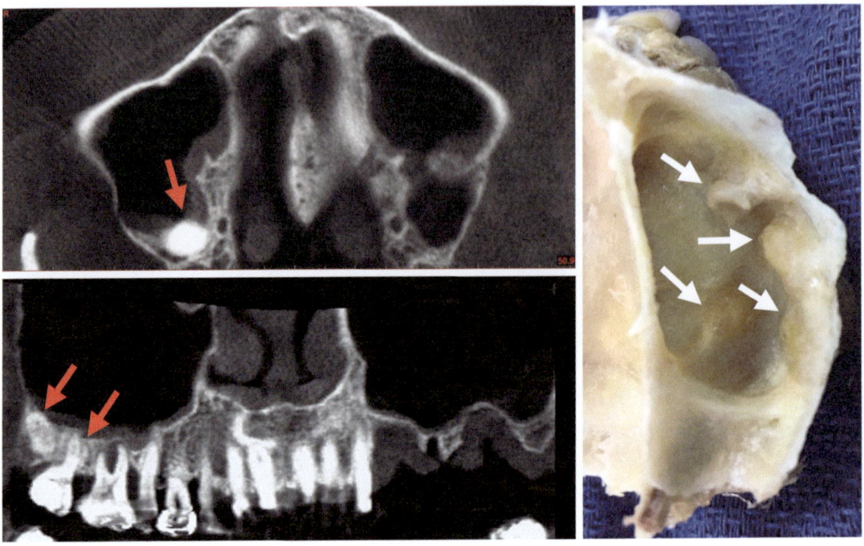

Fig. 8.2 Sinus floor close to the roots of the maxillary teeth

and sinusitis. Endodontic therapy or extraction of these teeth can result in penetration, oroantral fistulae, or root displacement into the sinus cavity. A periapical or periodontal infection of the upper premolars and molars can spread beyond the confines of the supporting dental tissue into the maxillary sinus, causing sinusitis (Hauman et al. 2002). The maxillary sinus with its cortical bone layer is a natural barrier against intrusion of the maxillary molars. Intrusion or bodily movement of the posterior teeth across the maxillary sinus has been known to cause moderate apical root resorption and more tipping (Al-Faraje et al. 2013).

8.7 Septum

Bony septa that partially divide the sinus into two or more compartments are common (Fig. 8.3). Such septa could be formed during the different phases of sinus pneumatization. Increased antral pneumatization starting after tooth loss seems to result from basal bone loss caused by the osteoclastic activity of the sinus membrane. Septa were first described by the anatomist Underwood in 1910 (Underwood 1910). Their shape has been described as resembling an inverted gothic arch, arising from the inferior and lateral walls of the sinus, and coming to a sharp edge along its most apical border; however, there are variations in shape, size, and location (McGowan et al. 1993). More variations of the maxillary sinus septa are described in the literature, such as partial perpendicular and partial horizontal septa, and complete septation of the sinus by a complete vertical septum.

An analysis of 312 sinuses in 156 patients revealed septa in 24% of the sinuses and 33% of the patients (Velásquez-Plata et al. 2002). Another study reported on 200 sinuses in 100 patients and found 53 sinuses (26.5%) to contain one or more septa (Kim et al. 2006). In an edentulous population, the incidence of septa can be

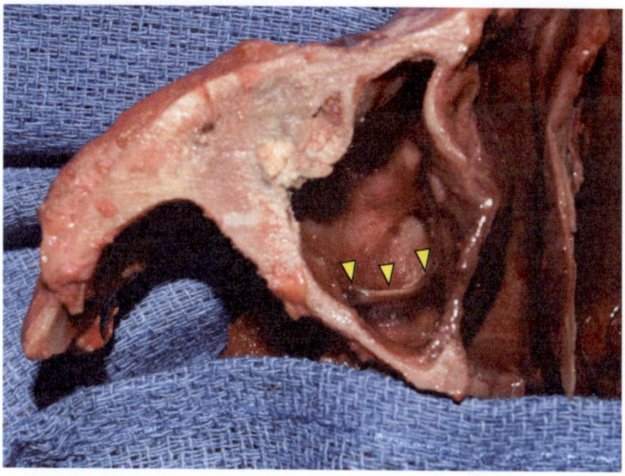

Fig. 8.3 Partial perpendicular septa

as high as 32% (Ulm et al. 1995). Maxillary septa can arise from the anterior, middle, or posterior portions of the sinus floor but most commonly are found in the middle third.

8.7.1 Partial Perpendicular Septa

Underwood's septa are osseous ridges delimiting the root compartments of the premolars and molars (Ulm et al. 1995; Cakur et al. 2013). They are encountered in 31–48% of cases and appear to develop in one of two ways, either primarily (i.e., from the development of the maxilla) or secondarily (i.e., as a result of tooth loss and remnant interseptal bone). Tooth loss and pneumatization adjacent to either a primary or a secondary septum can also exaggerate the height or size of the septum. It might be necessary to remove these septa during sinus grafting procedures as they can impede the view of the sinus floor and also limit placement of autogenous bone grafts or bone substitutes, thus preventing adequate filling of the sinus floor. Ulm et al. (1995) found no correlation between the six residual ridge classes described by Cawood and Howell and the incidence of Underwood's septa.

The punctum convergii is an area immediately posterior to the maxillary sinus ostium. In rare cases a partial perpendicular septum can develop into the maxillary sinus, beginning at this location, dividing the sinus into two incomplete compartments.

8.7.2 Partial Horizontal Septa

This septum is a horizontally located bony ridge. If it occurs, neither process will follow the normal development of the sinus in an inferior-medial direction. It will maintain their primary horizontal positions while the floor of the sinus is gradually lowered as it develops.

This type of septum can be ignored if it is positioned much higher than the area to be grafted. Otherwise, it will impede sinus drainage and lead to sinus graft failure.

8.7.3 Complete Septation of the Maxillary Sinus

A complete vertical septum usually divides the sinus into a large anterior and a smaller (accessory) posterior sinus. The anterior sinus drains into the middle meatus while the posterior sinus opens into the superior meatus through a bony hiatus delimited by structures similar to those of the normal hiatus semilunaris. Each of these sinuses can show isolated signs of pathology, and they must therefore be treated separately.

8.7.4 Surgical Importance in Oral Implantology

Septa can complicate the creation of an implant socket at the transalveolar approach and the bony window at the lateral approach and also increase the risk of perforation of the sinus membrane during its elevation. It is therefore important to identify these septa on CBCT scans and simulation software prior to surgery. Regular panoramic radiographs are insufficient to confirm the 3D structure of the septa and any pathological condition in the maxillary sinus. During sinus graft surgery, a septum can be left intact by making two windows separated by it, or the position of the implant should be planned as far as possible from it (Al-Faraje et al. 2013).

8.8 Sinus Floor Elevation

The vertical distance between the alveolar ridge and sinus floor is often insufficient for implant placement because of the bone resorption after tooth extraction and pneumatization in the molar region of maxilla (Fig. 8.4). Any pathological feature seen in the maxillary sinus has to be treated before the operation on the sinus. The molar teeth often become infected, and odontogenic sinusitis can develop; this can be treated by extraction of the affected teeth (Fig. 8.5). When sinusitis is not related to the oral cavity, the patient must be referred to an otorhinolaryngologist. After CBCT evaluation, according to the distance between the alveolar ridge and sinus floor or the inclination of the sinus floor, the treatment plan for implant placement and bone augmentation is determined. Two procedures can be used for sinus floor elevation depending on the area of approach into the sinus, either transalveolar or lateral. The decision tree is shown in Fig. 8.6.

8.8.1 Transalveolar Approach

If the bone height is more than 5 mm but it is difficult to place a regular implant in its length, the transalveolar approach is indicated (Pjetursson and Lang 2014). This procedure is limited to cases where the vertical augmentation is within 5 mm. The risk of perforation of the sinus floor membrane is reported to be greater if the distance of elevation exceeds 5 mm.

In this approach, the initial osteotomy can be created as a regular implant placement after careful preoperative measurement of the subsinus residual height. Osteotomy using drill is continued until the residual bone height beneath the

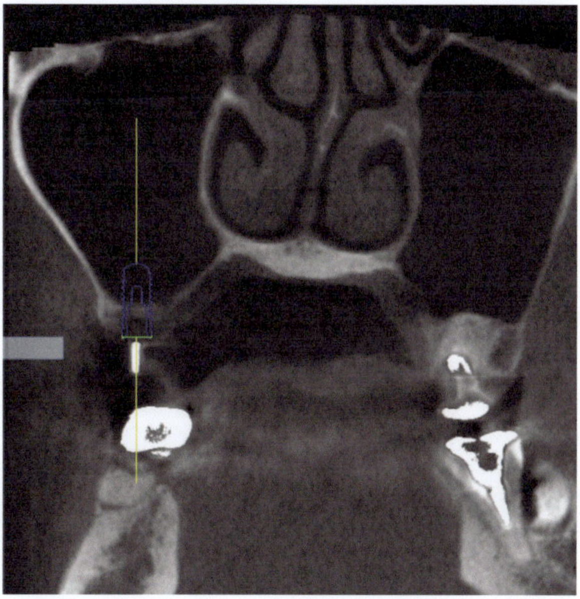

Fig. 8.4 Alveolar bone resorption and pneumatization after tooth extraction

maxillary sinus floor is limited to about 2 mm and the diameter of the socket according to the bone quality. Then an osteotome of the same size as the previous drilling is inserted into the implant socket, and the sinus floor is elevated by gentle malleting, resulting in a greenstick fracture. Once the osteotome reaches the planned implant length, bone substitutes are inserted from the sockets and pushed up by malleting one to three times. Finally, an implant is placed in the elevated socket.

One drawback of this procedure is that it is not known during the surgery whether the sinus floor membrane is perforated. Some studies have reported that when the sinus floor is not flat to the axial direction of implant, the incidence of sinus membrane perforation is greater.

8.8.2 Lateral Approach

If the distance from the alveolar ridge to the sinus floor is less than 5 mm, or the vertical augmentation would be more than 5 mm, the lateral approach should be selected. If the bone height is >5 mm but the sinus floor is inclined to the implant axis, this approach is indicated.

Furthermore, while transalveolar approach assumes simultaneous implant placement, whether simultaneous implant placement or only bone augmentation prior to implant placement is selected depends on the residual original bone height and on whether primary stability can be achieved as a result. Thus, the border is 3 mm of the original bone height.

The procedure is initiated by creating a bone window at the buccal side of the maxillary sinus using a sinus floor elevator after full-thickness flap elevation. Gentle elevation is needed so the membrane is not perforated. After elevation up to a distance sufficient for the planned implant placement, a resorbable membrane such as

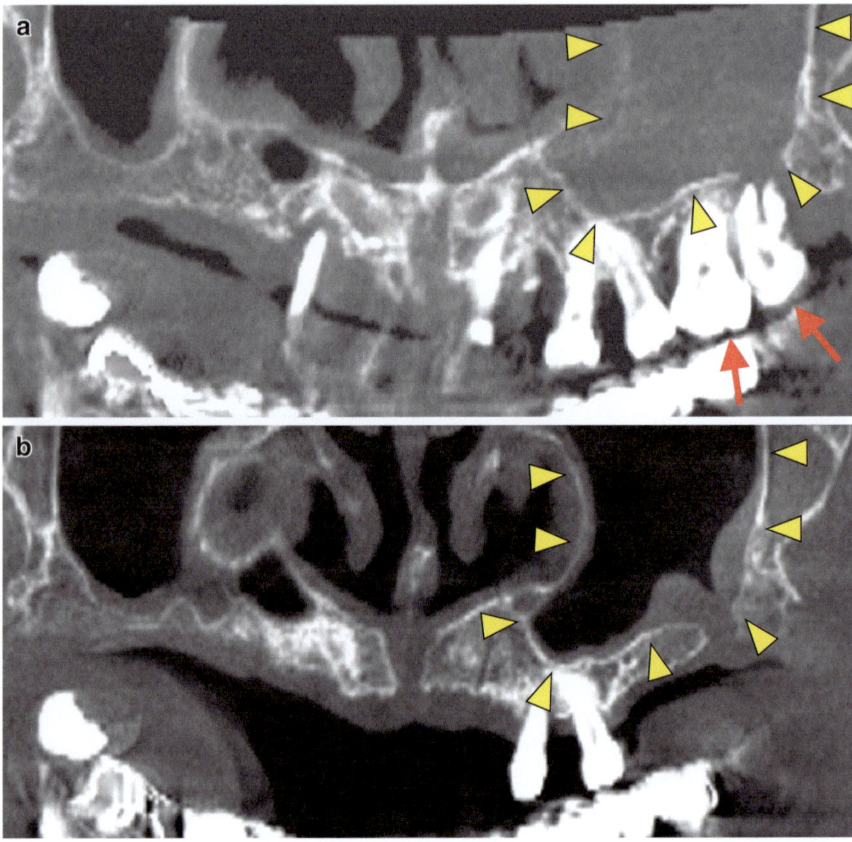

Fig. 8.5 Odontogenic sinusitis. (**a**) Before tooth extraction. (**b**) After tooth extraction

Fig. 8.6 Decision tree of sinus floor elevation

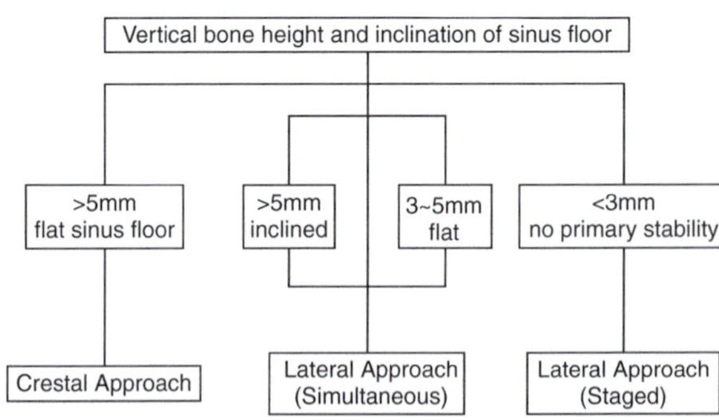

a collagen membrane is set at the top of the elevated space. The osteotomy is then performed and the implants are placed. Bone substitutes are embedded in the space around the implant body. Finally, the bone window is covered with resorbable membrane, and the flap is sutured closed (Fig. 8.7). If the bone height is less than 3 mm or primary stability cannot be achieved, implant placement should be planned after 6 months as a staged approach (Fig. 8.8).

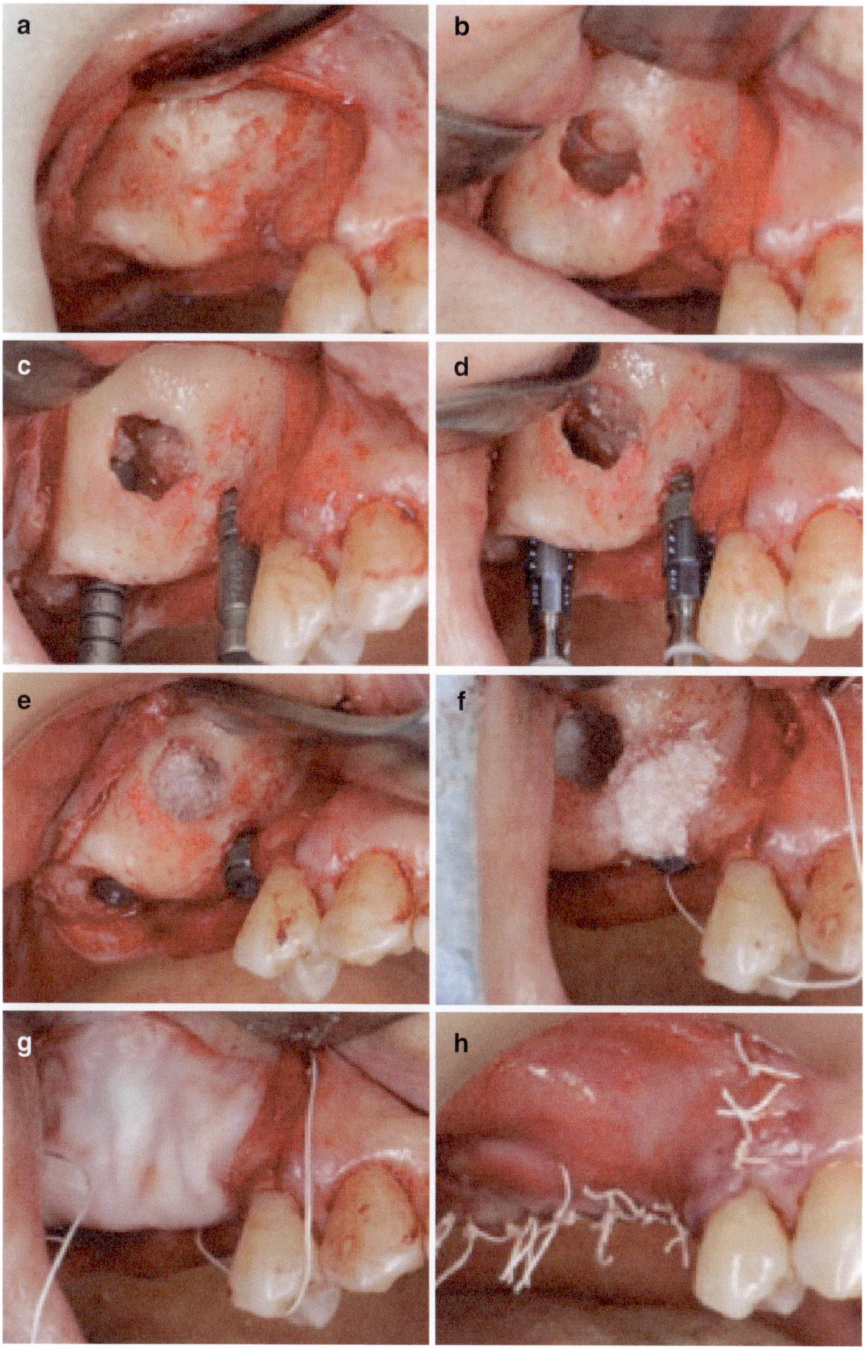

Fig. 8.7 Lateral approach and simultaneous implant placement with GBR

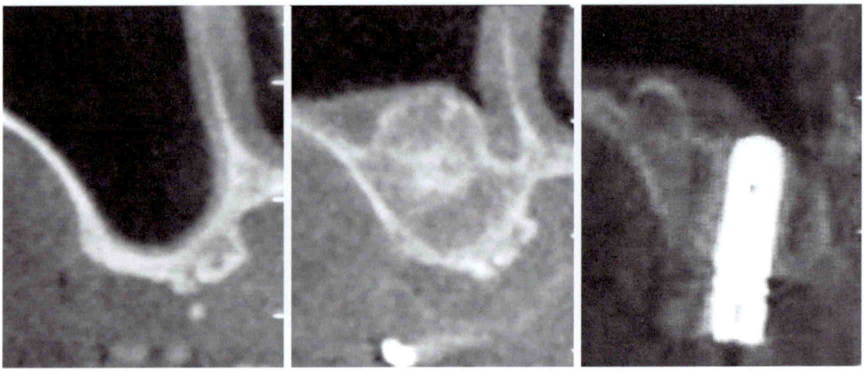

Fig. 8.8 Lateral approach and staged implant placement after 6 months

References

Al-Faraje L, Church C, Rathburn A (2013) Surgical and radiologic anatomy for oral implantology. Quintessence Publishing, Hanover Park, IL

Bailey BJ, Johnson JT, Newlands SD (2006) Head & neck surgery—otolaryngology. Lippincott Williams & Wilkins, Philadelphia, PA

Cakur B, Sümbüllü MA, Durna D (2013) Relationship among Schneiderian membrane, Underwood's septa, and the maxillary sinus inferior border. Clin Implant Dent Relat Res 15:83–87

Eggesbø HB (2006) Radiological imaging of inflammatory lesions in the nasal cavity and paranasal sinuses. Euro Radiol 16:872–888

Hauman CHJ, Chandler NP, Tong DC (2002) Endodontic implications of the maxillary sinus: a review. Int Endod J 35:127–141

Keir J (2009) Why do we have paranasal sinuses? J Laryngol Otol 123:4–8

Kim M-J, Jung U-W, Kim C-S et al (2006) Maxillary sinus septa: prevalence, height, location, and morphology. A reformatted computed tomography scan analysis. J Periodontol 77:903–908. https://doi.org/10.1902/jop.2006.050247

Lorkiewicz-Muszyńska D, Kociemba W, Rewekant A et al (2015) Development of the maxillary sinus from birth to age 18. Postnatal growth pattern. Int J Pediatr Otorhinolaryngol 79:1393–1400

McGowan DA, Baxter PW, James J (1993) The maxillary sinus and its dental implications. John Wright, Oxford

Pjetursson BE, Lang NP (2014) Sinus floor elevation utilizing the transalveolar approach. Periodontol 2000 66:59–71

Ulm CW, Solar P, Krennmair G et al (1995) Incidence and suggested surgical management of septa in sinus-lift procedures. Int J Oral Maxillofac Implants 10:462–465

Underwood AS (1910) An inquiry into the anatomy and pathology of the maxillary sinus. J Anat Physiol 44:354–369

van den Bergh JP, ten Bruggenkate CM, Disch FJ et al (2000) Anatomical aspects of sinus floor elevations. Clin Oral Impl Res 11:256–265

Velásquez-Plata D, Hovey LR, Peach CC et al (2002) Maxillary sinus septa: a 3-dimensional computerized tomographic scan analysis. Int J Oral Maxillofac Implants 17(6):854–860

Wehrbein H, Diedrich P (1992) The initial morphological state in the basally pneumatized maxillary sinusDOUBLEHYPHENa radiological-histological study in man. Fortschr Kieferorthop 53:254–262

Anatomy and Variations of the Posterior Superior Alveolar Artery and Nerve

9

Iwona M. Tomaszewska, Patrick Popieluszko, Krzysztof A. Tomaszewski, and Jerzy A. Walocha

9.1 Introduction

The posterior superior alveolar artery is a terminal branch of the maxillary artery supplying the maxillary sinus, the upper molars and premolars, and parts of the buccal gingiva (Solar et al. 1999). As this artery enters the maxilla from the posterior superior alveolar foramen, it runs with the posterior superior alveolar branch of the maxillary nerve that innervates the upper molars (Kasahara et al. 2016; Iwanaga et al. 2017). Their course through the maxillary sinus and surface of the maxilla can vary markedly, and this should be kept in mind during operations in this area. Their anatomy is especially important for dental procedures, e.g., maxillary implant placement, sinus floor augmentation, anesthesia of the maxillary molars, Le Fort I osteotomy, and Le Fort I fracture repairs. In this chapter we describe the commonly encountered variants in the anatomy of the posterior superior alveolar nerve and artery and the superior alveolar canals associated with the neurovascular bundle and discuss the clinical significance of these structures and the potential effects of variants on clinical practice.

9.2 Posterior Superior Alveolar Artery and its Variations

As the maxillary artery crosses into the maxillary sinus through its posterior wall, it gives off a branch known as the posterior superior alveolar artery (PSAA) that travels along the wall of the sinus, supplying the mucosa, the maxillary molar teeth, and parts of the buccal gingiva (Solar et al. 1999; Hur et al. 2009) (Fig. 9.1). As it travels

I. M. Tomaszewska (✉) · P. Popieluszko · K. A. Tomaszewski · J. A. Walocha
Department of Anatomy, Jagiellonian University Medical College, Krakow, Poland

J. Iwanaga, R. S. Tubbs (eds.), *Anatomical Variations in Clinical Dentistry*,
https://doi.org/10.1007/978-3-319-97961-8_9

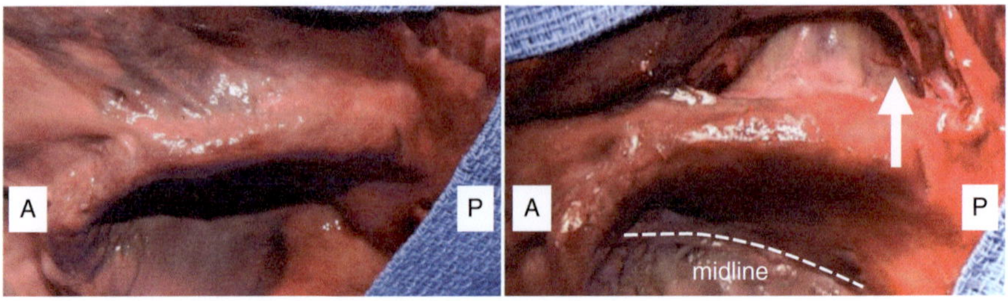

Fig. 9.1 Left posterior superior alveolar artery (arrow)

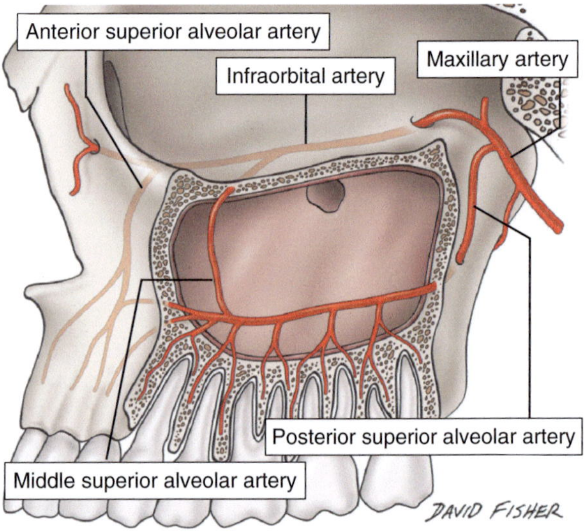

Fig. 9.2 Anastomosis between PSAA and infraorbital artery

through the sinus, it provides an interosseous and an extraosseous branch, the final result being terminal anastomosis between the PSAA and the branches of the infraorbital artery (Güncü et al. 2011) (Fig. 9.2). Most variations of the PSAA involve its path through the sinus and bony structures. The PSAA most often travels within the maxilla (Fig. 9.3), but it has been noted in the mucosa of the sinus and in rare cases superficial to the mucosa or external to the lateral wall of the maxillary sinus. The reported prevalence of these patterns across studies is summarized in Table 9.1.

It is worth noting that cone beam computed tomography (CBCT) could identify the artery in only 87 and 60.58% of the populations studied, the PSAA being more difficult to detect in edentulous patients (Danesh-Sani et al. 2017; Tehranchi et al. 2017). There is a statistically significant difference between males and females, males having an interosseous path more often than females (Tehranchi et al. 2017). The course of the PSAA from the origin at the maxillary artery to the point where it anastomoses with the infraorbital artery can also be characterized as one of two patterns: straight or U-shaped (Hur et al. 2009). The straight path is seen in the vast

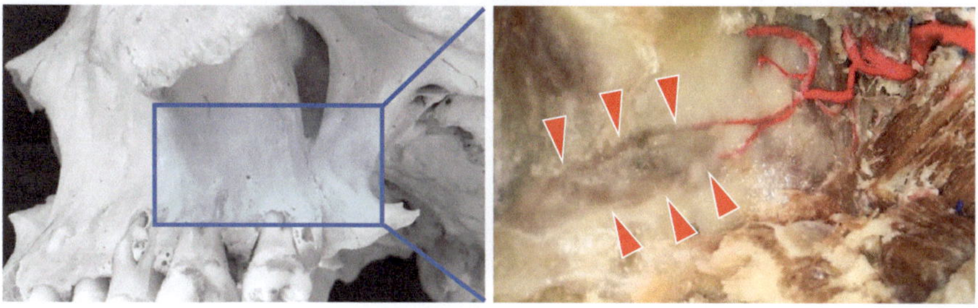

Fig. 9.3 Course of PSAA

Table 9.1 Posterior superior alveolar artery course relative to maxillary bone structure

Study	Intraosseous (%)	Intrasinusal (%)	Superficial/external (%)
Tehranchi et al. (2017)	47	47	6 (external)
Danesh-Sani et al. (2017)	69.6	24.3	6.1
Güncü et al. (2011)	68.2	26	5.7
Ella et al. (2008)	71.4	14.3	N/A
Ilguy et al. (2013)	71.1	13	5.2
Kang et al. (2013)	64.3	29.1	6.6

Table 9.2 Distance from PSAA to maxillary sinus floor (mm)

Study	1st premolar	2nd premolar	1st molar	2nd molar
Danesh-Sani et al. (2017)	8.57	8.45	8.09	9.27
Yang and Kye (2014)	6.67	8.19	7.59	7.8
Watanabe et al. (2014)	10.4	10.4	9.0	8.5
Apostolakis and Bissoon (2014)	5.9	5.8	5.9	8.0
Jung et al. (2011)	9.2	8.07	7.58	9.02
Hur et al. (2009)	9.4	9.7	10.3	9.6

majority of subjects, with prevalence as high as 78.1 and 80% in the populations studied (Hur et al. 2009; Santos German et al. 2015).

Measurements of the PSAA relative to other structures differ from study to study. They include distances of the PSAA from the zygomatic arch, the nasal septum, the medial wall of the maxillary sinus, the alveolar crest, and the maxillary sinus floor or the cementoenamel junction of the maxillary teeth. These measurements also can vary along the length of the PSAA. The most commonly measured and comparable distances are summarized in Tables 9.2 and 9.3.

Various distances were measured depending on the study design, and there has been little uniformity across studies except for measurements from the floor. In contrast, diameter measurements have been rather uniform across studies, as summarized in Table 9.4. Furthermore, no statistically significant difference has been noted between age groups, but men have larger-lumened PSAAs than women (Güncü et al. 2011; Tehranchi et al. 2017).

Table 9.3 Distance from PSAA to related structures (mm)

Study	Alveolar crest	Zygomatic arch	Nasal septum	Medial wall of maxillary sinus
Tehranchi et al. (2017)	16.7 ± 3.96	25.59 ± 4.89	26.51 ± 3.52	
Güncü et al. (2011)	18.0 ± 4.9			11.0 ± 3.8
Watanabe et al. (2014)	19.0			

Table 9.4 Reported diameters of the PSAA

Study	Diameter (mm)
Ella et al. (2008)	1.20 ± 0.5
Tehranchi et al. (2017)	1.29 ± 0.36
Güncü et al. (2011)	1.3 ± 0.5
Danesh-Sani et al. (2017)	1.17 ± 0.6
Ilguy et al. (2013)	0.94 ± 0.26
Kang et al. (2013)	1.21 ± 0.55
Hur et al. (2009)	0.9 ± 0.38

9.3 Posterior Superior Alveolar Nerve

The posterior superior alveolar nerve (PSAN) is one of three superior alveolar branches. It arises from the maxillary nerve that travels with the PSAA (Solar et al. 1999) (Fig. 9.4). Before it splits into small fibers, the observable portion of the PSAN travels parallel to and with the PSAA (Sato et al. 2010) (Fig. 9.5). Its fibers are particularly important as histological evidence shows them to contain calcitonin gene-related peptide. They are nociceptive and influence the vascular tone of the PSAA along with sensation to the alveolar structure of the maxilla (Sato et al. 2010). The anatomical variants of this nerve and its branches have not been studied extensively, despite its importance; much more time has been invested in the study of the anatomy surrounding the PSAN in order to enable dentists and surgeons to target it properly for anesthesia. The most common way to access the PSAN for blockade is to insert the needle with the anesthetic between the buccal mucosa and the alveolar process of the maxilla, at the mucobuccal fold in the area of the molar to be treated (Maljaei et al. 2017).

9.4 Posterior Superior Alveolar Canals

The PSAA and PSAN can vary in their course through the sinus and bone, but the posterior superior alveolar canals should be considered since most populations studied have a PSAA/PSAN with some intraosseous involvement. The canal through which the PSAA and PSAN travel after entering the maxillary sinus has been described in three variations: canal, groove or ditch shaped, and fragmented (or partially intraosseous and partially extraosseous) (Sato et al. 2010) (Figs. 9.6 and 9.7). In the original

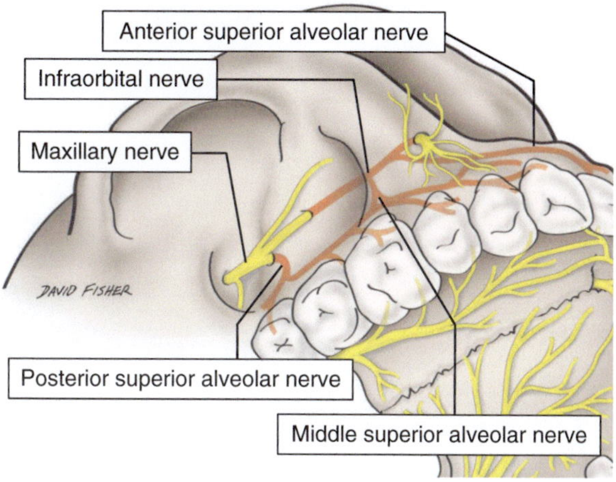

Fig. 9.4 Posterior, middle, and anterior superior alveolar nerves

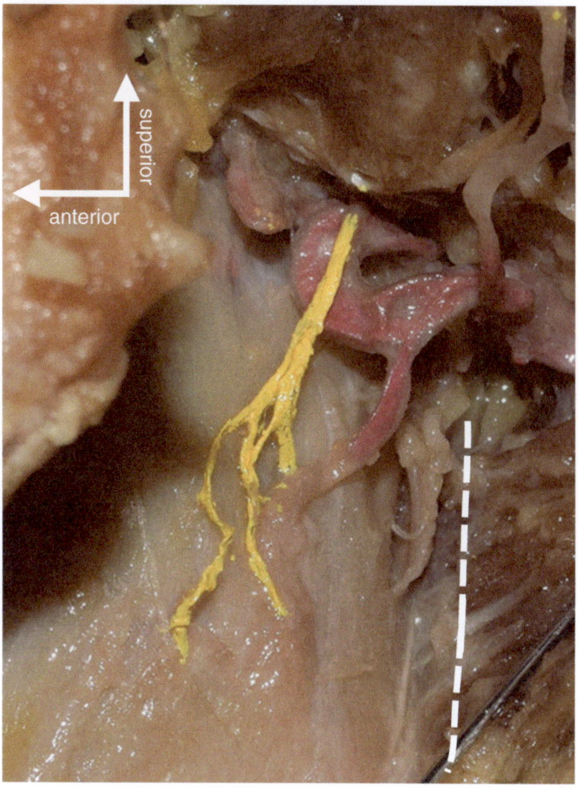

Fig. 9.5 Left posterior superior alveolar nerve (yellow) and artery (red). Dotted line; posterior border of the maxilla

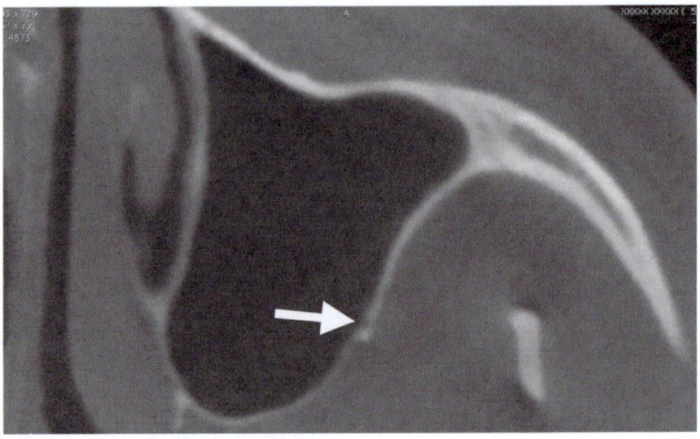

Fig. 9.6 Posterior superior alveolar canal (arrow) on axial CT

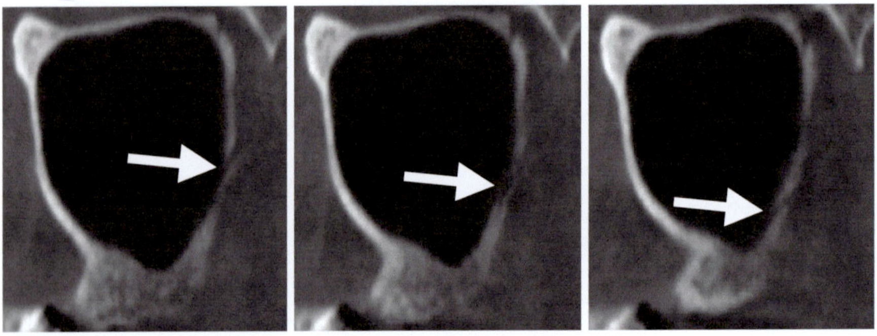

Fig. 9.7 Posterior superior alveolar canal (arrows) in sagittal CT

study describing these shapes, the groove or ditch shape was the most common, seen in 67.6%. A full canal was only seen in 14.7% of cases, and the fragmented variant was observed in 17.6% (Sato et al. 2010). A subsequent study, using various modalities including 90 dry skulls, revealed similar findings; 65% of the skulls showed the groove presentation (Santos German et al. 2015). One CBCT scan study focused on the canal itself, when it was present, reporting a mean diameter of 1.1 ± 0.4 mm with no statistically significant differences between males and females (Apostolakis and Bissoon 2014). As with the reported diameters of the PSAA, there were no significant age differences in canal diameter (Apostolakis and Bissoon 2014).

9.5 Clinical Implication

Poor comprehension of the PSAA, the bony structure of the maxilla, and the PSAN and their respective anatomies can lead to disastrous complications during several surgical procedures. Therefore it is imperative that surgeons operating in this area of the face understand the possible variants they could encounter and prepare accordingly. The procedures where this consideration is most important include maxillary dental implants, sinus floor elevation/augmentation, Le Fort I osteotomy, Le Fort I fracture repair, maxillary sinus surgery, and anesthesia of the molars for dental and periodontal procedures.

9.5.1 Implant surgery

For implants in the maxilla, the course of the PSAA and the associated PSAN fibers must be considered to avoid injury, excessive hemorrhage, and postoperative pain caused by inflammation (Peñarrocha-Diago et al. 2009; Sato et al. 2010). The most common path of the PSAA through the maxilla, the groove type, could be more susceptible to injury as the walls are thinner (Sato et al. 2010). This puts most patients at higher risk of untoward injury to the PSAA. Furthermore, as inflammation in the periapical area of the implant is most readily treated by further surgery, it is prudent to maximize the prevention of such complications through proper understanding of each patient's individual anatomy (Peñarrocha-Diago et al. 2009).

Aside from the complications that can arise during the actual implant placement, the PSAA should also be considered in any procedure performed to help prepare the site for implantation. This includes procedures such as sinus floor elevation and augmentation of the maxilla, which serve to create more bone stock for implant placement and surgical success (Tong et al. 1998). Not only can the PSAA be a source of unwanted bleeding during the procedure, but it is also crucial for proper vascularization of the graft material (Traxler et al. 1999; Ella et al. 2008). Excessive hemorrhage from the PSAA is less common with augmentation. In this case, if the artery has a smaller diameter, there is less risk of encountering it in the operating field (Güncü et al. 2011). However, the lift procedure carries the risk of complete transection of the artery and hemorrhage, regardless of diameter (Greenstein et al. 2008). As there might be alveolar canals, which could obstruct the Caldwell-Luc antrostomy, the PSAA is considered a possible source of bleeding when hemostasis is hard to achieve (Rahpeyma and Khajehahmadi 2014). A modified Caldwell-Luc approach has been suggested, the window being created 2–3 mm above the floor of the maxillary sinus, which promotes healing and remodeling and also stability of the graft material (Hur et al. 2009).

9.5.2 Le Fort I fracture and osteotomy

When a patient with a Le Fort I fracture is treated, the PSAA should be considered in cases of excessive hemorrhage. The position of the osteotomy line in such a repair has been found to coincide with the position of the PSAA (Hur et al. 2009). A case in which a patient presented with a Le Fort I fracture has been described; postoperatively, there was uncontrolled oronasal bleeding leading to hemodynamic instability despite transfusion (Hwang and Choi 2009). Only after angiography was the PSAA identified as the source of bleeding, and prompt embolization provided

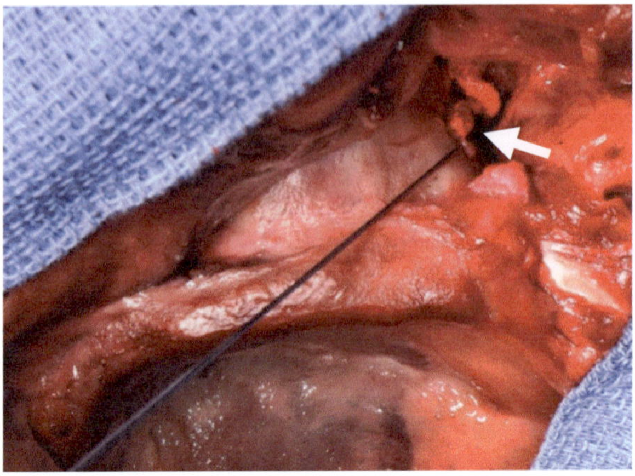

Fig. 9.8 Needle approaching the maxillary artery (arrow) and the PSAA and PSAN

relief (Hwang and Choi 2009). It was therefore suggested that angiography should be considered for visualization of the PSAA prior to surgery if a patient with a Le Fort fracture presents with hemodynamic instability (Hwang and Choi 2009). Considerations have also been extended to Le Fort I elective osteotomies: in all cases of such fractures, it is important to ensure that the PSAA and its canals are not mistaken for a fracture line (Rahpeyma and Khajehahmadi 2014).

9.5.3 Posterior superior alveolar nerve block

The PSAN is a common target for nerve block when procedures involving the upper molars are considered, though it is reported to be no more effective than buccal infiltration and combined buccal and palatal infiltration (Aggarwal et al. 2011). Aside from effectiveness, certain risks should be considered with all local anesthesia. For PSAN block specifically, the most common risk is hematoma at the injection site (Fig. 9.8), but there can be more serious consequences, even ocular impairment (Harn et al. 2002; Ghosh et al. 2015). There have been many suggestions about why local injection can cause ocular symptoms, including infiltration of the PSAA, which is nearby (Lee 2006). In the search for an ideal and safe place, an area over the second maxillary molar has been described as providing adequate anesthesia in 99% of subjects studied while lacking neurovascular tissue for possible infiltration (Harn et al. 2002). To make access to the PSAN easier, it has also been suggested that patients have their mandibles shifted toward the side of the injection to provide more room for visualization of the target area, especially in wisdom tooth extraction (Iwanaga et al. 2017) (Fig. 9.9). As the fibers of the PSAN can be difficult to visualize on imaging owing to their size and the contrast of the PSAA overshadowing them, studies have instead focused on the canal as a whole when CBCT is used for visualization (Apostolakis and Bissoon 2014).

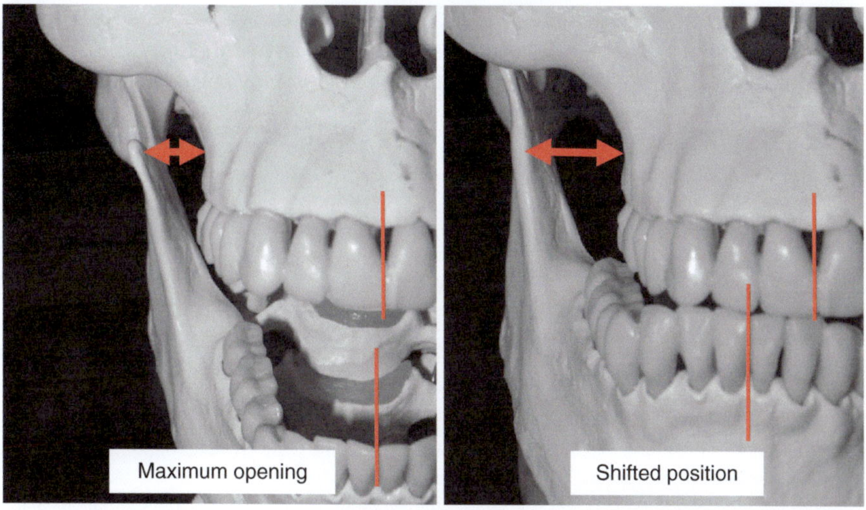

Fig. 9.9 Shifting mandible to ipsilateral side to provide more room for block procedure and wisdom tooth extraction

9.5.4 Anatomical variation in size

In all cases of surgery close to the PSAA, both the course relative to structures and the diameter of the vessel are of paramount importance. The distance of the PSAA from the sinus floor has been readily measured in multiple studies. However, it has been proved highly variable, so the mean values described should not serve as exclusive guidelines on which to base clinical practice (Danesh-Sani et al. 2017). The only suggestion offered on the basis of these measurements has been extra caution at the level of the first molar, where the vessel height from the floor is lower than at the level of the first premolar (Hur et al. 2009). Furthermore, a patient's dental status can complicate the expected localization. The distance between the canal and its structures and the sinus floor is greater in edentulous patients (Apostolakis and Bissoon 2014). In some studies, it was proved the PSAA is more difficult to be visualized in edentulous patients, leading to a lower rate of detection of the PSAA on imaging (Tehranchi et al. 2017). The diameter of the PSAA in a patient serves as a better guideline for estimating the risk of bleeding. It has been suggested that diameters >2 mm entail the greatest risk of excessive bleeding (Apostolakis and Bissoon 2014). Although vessels >2 mm are rare, the area in which a portion of the PSAA of this thickness is most commonly found is at the point of the posterior superior alveolar foramen (Hur et al. 2009). The lack of correlation with age or laterality leaves only a suggestion for extra caution in male patients, who routinely have larger diameter arteries and are therefore at higher risk of untoward hemorrhage.

In preparation for surgeries in the area of the maxilla and the maxillary sinus, including implant placement, sinus floor elevation and augmentation, Le Fort I fracture repair, and local anesthesia of the maxillary molars, consideration of the PSAA and its associated structures is critical for a successful procedure. Since many factors influence the position of the PSAA, the canal, and its associated neural fibers, it is recommended that CBCT be performed prior to any procedure concerning the maxilla and the maxillary sinus (Sato et al. 2010; Apostolakis and Bissoon 2014; Danesh-Sani et al. 2017).

References

Aggarwal V, Singla M, Miglani S et al (2011) A prospective, randomized, single-blind comparative evaluation of anesthetic efficacy of posterior superior alveolar nerve blocks, buccal infiltrations, and buccal plus palatal infiltrations in patients with irreversible pulpitis. J Endod 37:1491–1494. https://doi.org/10.1016/J.JOEN.2011.08.018

Apostolakis D, Bissoon AK (2014) Radiographic evaluation of the superior alveolar canal: measurements of its diameter and of its position in relation to the maxillary sinus floor: a cone beam computerized tomography study. Clin Oral Implants Res 25:553–559. https://doi.org/10.1111/clr.12119

Danesh-Sani SA, Movahed A, ElChaar ES et al (2017) Radiographic evaluation of maxillary sinus lateral wall and posterior superior alveolar artery anatomy: a cone-beam computed tomographic study. Clin Implant Dent Relat Res 19:151–160. https://doi.org/10.1111/cid.12426

Ella B, Sédarat C, Noble RDC et al (2008) Vascular connections of the lateral wall of the sinus: surgical effect in sinus augmentation. Int J Oral Maxillofac Implants 23:1047–1052

Ghosh A, Vaibhav N, Raut R et al (2015) Ophthalmic complication following posterior superior alveolar nerve block for tooth extraction! A rare occurrence. J Maxillofac Oral Surg 14:862–865. https://doi.org/10.1007/s12663-015-0756-7

Greenstein G, Cavallaro J, Romanos G et al (2008) Clinical recommendations for avoiding and managing surgical complications associated with implant dentistry: a review. J Periodontol 79:1317–1329. https://doi.org/10.1902/jop.2008.070067

Güncü GN, Yildirim YD, Wang H-L et al (2011) Location of posterior superior alveolar artery and evaluation of maxillary sinus anatomy with computerized tomography: a clinical study. Clin Oral Implants Res 22:1164–1167. https://doi.org/10.1111/j.1600-0501.2010.02071.x

Harn SD, Durham TM, Callahan BP et al (2002) The triangle of safety: a modified posterior superior alveolar injection technique based on the anatomy of the PSA artery. Gen Dent 50:554-7-9

Hur M-S, Kim J-K, Hu K-S et al (2009) Clinical implications of the topography and distribution of the posterior superior alveolar artery. J Craniofac Surg 20:551–554. https://doi.org/10.1097/SCS.0b013e31819ba1c1

Hwang K, Choi HG (2009) Bleeding from posterior superior alveolar artery in Le fort I fracture. J Craniofac Surg 20:1610–1612. https://doi.org/10.1097/SCS.0b013e3181b14775

Ilguy D, Ilguy M, Dolekoglu S et al (2013) Evaluation of the posterior superior alveolar artery and the maxillary sinus with CBCT. Braz Oral Res 27:431–437. https://doi.org/10.1590/S1806-83242013000500007

Iwanaga J, Simonds E, Oskouian RJ et al (2017) Cadaveric study for intraoral needle access to the infratemporal fossa: application to posterior superior alveolar nerve block technique. Cureus 9:e1761. https://doi.org/10.7759/cureus.1761

Jung J, Yim J-H, Kwon Y-D et al (2011) A radiographic study of the position and prevalence of the maxillary arterial endosseous anastomosis using cone beam computed tomography. Int J Oral Maxillofac Implants 26:1273–1278

Kang S-J, Shin S-I, Herr Y et al (2013) Anatomical structures in the maxillary sinus related to lateral sinus elevation: a cone beam computed tomographic analysis. Clin Oral Implants Res 24:75–81. https://doi.org/10.1111/j.1600-0501.2011.02378.x

Kasahara N, Morita W, Tanaka R et al (2016) The relationships of the maxillary sinus with the superior alveolar nerves and vessels as demonstrated by cone-beam CT combined with μ-CT and histological analyses. Anat Rec 299:669–678. https://doi.org/10.1002/ar.23327

Lee C (2006) Complications after inferior alveolar nerve block hypothesis of ocular complications. Hong Kong Med Diary: 11

Maljaei E, Pourkazemi M, Ghanizadeh M et al (2017) The efficacy of buccal infiltration of 4% articaine and PSA injection of 2% lidocaine on anesthesia of maxillary second molars. Iran Endod J 12:276–281. https://doi.org/10.22037/iej.v12i3.16464

Peñarrocha-Diago M, Boronat-Lopez A, García-Mira B (2009) Inflammatory implant periapical lesion: etiology, diagnosis, and treatment—presentation of 7 cases. J Oral Maxillofac Surg 67:168–173. https://doi.org/10.1016/j.joms.2007.12.022

Rahpeyma A, Khajehahmadi S (2014) Alveolar antral artery: review of surgical techniques involving this anatomic structure. Iran J Otorhinolaryngol 26:73–78

Santos German IJ, Buchaim DV, Andreo JC et al (2015) Identification of the bony canal of the posterior superior alveolar nerve and artery in the maxillary sinus: tomographic, radiographic, and macroscopic analyses. ScientificWorldJournal 2015:878205. https://doi.org/10.1155/2015/878205

Sato I, Kawai T, Yoshida S et al (2010) Observing the bony canal structure of the human maxillary sinus in Japanese cadavers using cone beam CT. Okajimas Folia Anat Jpn 87:123–128

Solar P, Geyerhofer U, Traxler H et al (1999) Blood supply to the maxillary sinus relevant to sinus floor elevation procedures. Clin Oral Implants Res 10:34–44

Tehranchi M, Taleghani F, Shahab S et al (2017) Prevalence and location of the posterior superior alveolar artery using cone-beam computed tomography. Imaging Sci Dent 47:39. https://doi.org/10.5624/isd.2017.47.1.39

Tong DC, Rioux K, Drangsholt M et al (1998) A review of survival rates for implants placed in grafted maxillary sinuses using meta-analysis. Int J Oral Maxillofac Implants 13:175–182

Traxler H, Windisch A, Geyerhofer U et al (1999) Arterial blood supply of the maxillary sinus. Clin Anat 12:417–421. https://doi.org/10.1002/(SICI)1098-2353(1999)12:6<417::AID-CA3>3.0.CO;2-W

Watanabe T, Shiota M, Gao S et al (2014) Verification of posterior superior alveolar artery distribution in lateral wall of maxillary sinus by location and defect pattern. Quintessence Int 45:673–678. https://doi.org/10.3290/j.qi.a32239

Yang S-M, Kye S-B (2014) Location of maxillary intraosseous vascular anastomosis based on the tooth position and height of the residual alveolar bone: computed tomographic analysis. J Periodontal Implant Sci 44:50. https://doi.org/10.5051/jpis.2014.44.2.50

Part IV

Hard Palate

Anatomy and Variations of the Greater Palatine Foramen

10

Iwona M. Tomaszewska, Patrick Popieluszko,
Krzysztof A. Tomaszewski, and Jerzy A. Walocha

10.1 Introduction

The greater palatine foramen (GPF) (Fig. 10.1) is one of the most important neurovascular conduits in the oral cavity (Fig. 10.2). It is formed between the maxilla and palatine bone (Fig. 10.3). The neurovascular bundle carries sensation and circulation to the gums, palate, and nasal cavity. Whether a surgeon is looking to prevent iatrogenic injury or aiming to provide hemostasis or anesthesia via this neurovasculature, knowledge of the exact location and possible anatomical variations of the GPF are vital for success. It is therefore particularly important for maxillofacial surgeries, periodontal procedures, local anesthesia, and hemostasis even beyond the scope of surgery. In this chapter, we review the normal anatomy of the foramen and the most common variants and discuss the clinical importance of the GPF and the associated greater palatine canal.

10.2 Anatomical Variations of the Greater Palatine Foramen, Canal, Nerve, and Artery

The greater palatine foramen (GPF) is the distal end point of the greater palatine canal (GPC), the connection between the oral cavity and the pterygopalatine fossa (Fig. 10.4). The canal and the foramen serve to carry the descending palatine artery, the greater palatine nerve, and the lesser palatine nerve, which provide the neurovascular supply for the palate and for some structures in the nasal cavity (Hwang et al. 2011) (Figs. 10.5 and 10.6). It is usually a single foramen, but it has been observed as a double foramen in 16% of studied skulls and rarely as a triple

I. M. Tomaszewska (✉) · P. Popieluszko · K. A. Tomaszewski · J. A. Walocha
Department of Anatomy, Jagiellonian University Medical College, Krakow, Poland

J. Iwanaga, R. S. Tubbs (eds.), *Anatomical Variations in Clinical Dentistry*,
https://doi.org/10.1007/978-3-319-97961-8_10

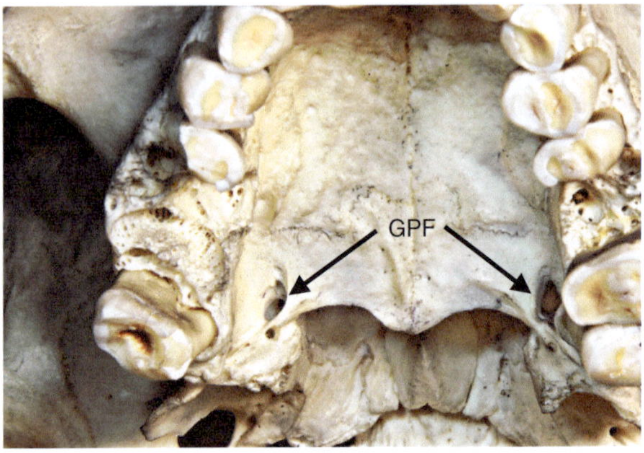

Fig. 10.1 Greater palatine foramina (GPF)

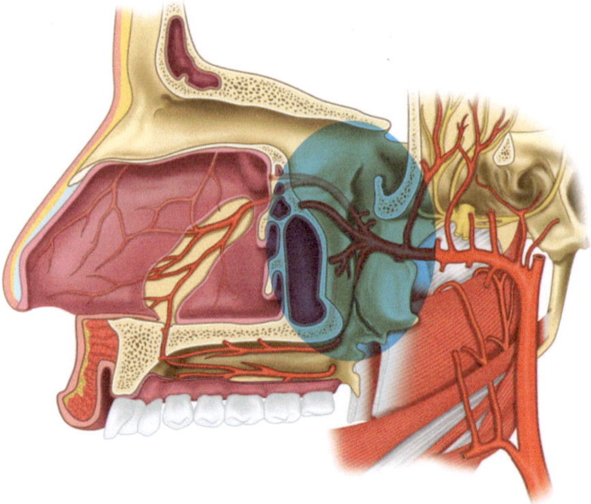

Fig. 10.2 Descending palatine artery and greater palatine nerve

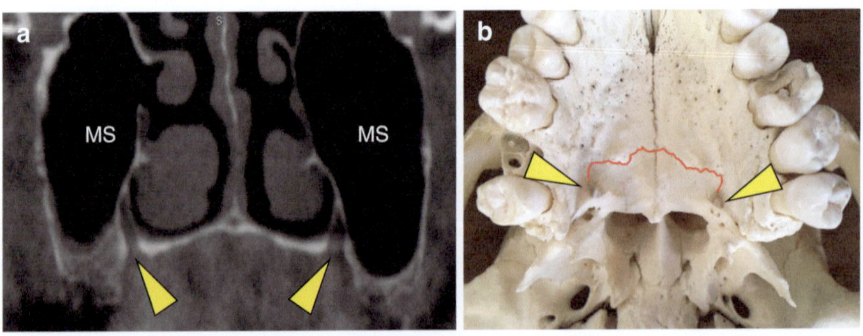

Fig. 10.3 Greater palatine foramina formed between maxilla and palatine bone. Note the foramina are on the border of these two bones (red line). (**a**) coronal CT image showing the hard palate and the sinuses. (**b**) bony specimen of the hard palate

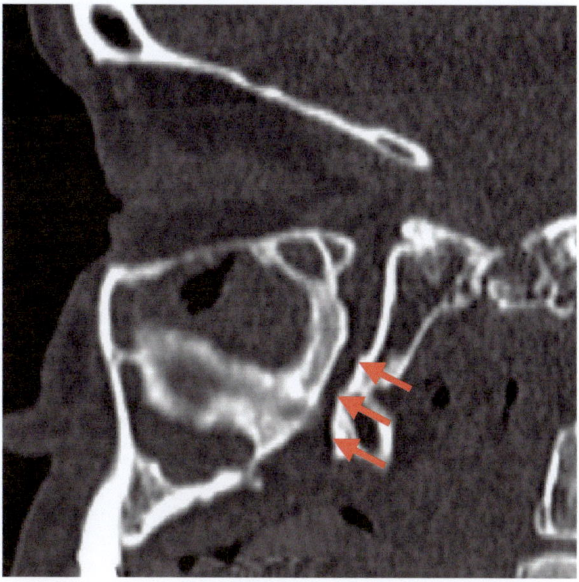

Fig. 10.4 Head CT (sagittal section) showing the pterygopalatine fossa and greater palatine canal (arrows)

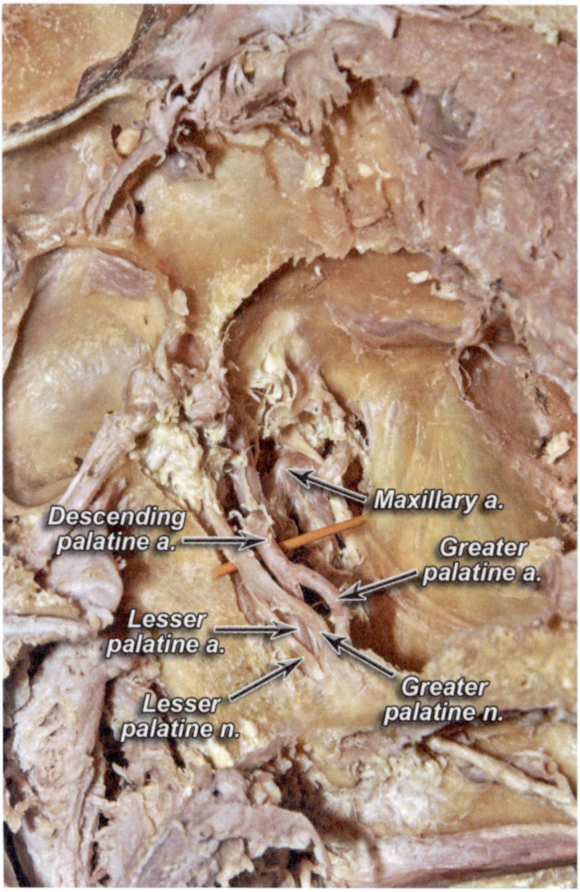

Fig. 10.5 Cadaveric specimen (head, sagittal section) showing the maxillary artery, descending palatine artery, greater palatine artery and nerve, lesser palatine artery and nerve

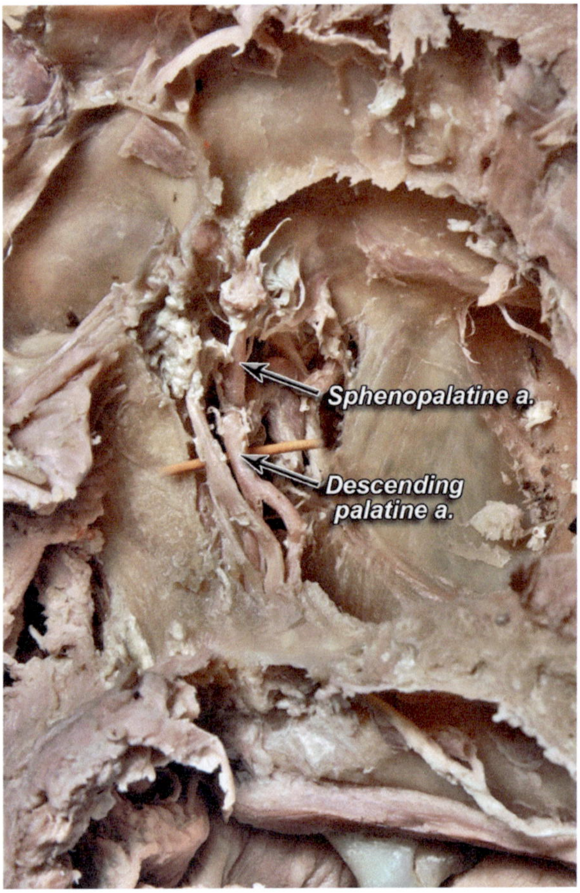

Fig. 10.6 Cadaveric specimen (head, sagittal section) showing the descending palatine artery and the sphenopalatine artery

foramen. It can even be absent in up to 2% of cases (Cagimni et al. 2017). The GPC usually traverses the palatine bone and opens as the GPF opposite the third molar (Howard-Swirzinski et al. 2010; Tomaszewska et al. 2014). Although the "normal" position of the GPF is considered to be at the level of the third molar, this has been shown to hold true in only 41.38–58.7% of subjects in the most recent studies (Aoun et al. 2015; Sarilita and Soames 2016). The GPF can be found anterior or posterior to this traditional position in the palate, and its second most common location is between the second and third molars, seen in 27.59–37.3% of subjects in recent studies (Aoun et al. 2015; Sarilita and Soames 2016). It rarely opens as anteriorly as the second molar; this has been seen in only 1.72–4.0% of subjects studied (Aoun et al. 2015; Sarilita and Soames 2016). The most recent studies on the position of the GPF relative to the molars are summarized in Table 10.1.

The GPF can also vary in its direction of opening relative to the palate. In relation to the GPC and the sagittal plane, it opens in an inferior-anterior-medial direction 82.1% of the time (Tomaszewska et al. 2015). The second most common direction is anterior, seen 7.6% of the time, with rare instances of vertical and

Table 10.1 Position of the GPF in relation to molars across studies

	GPF location relative to second molar (M2) and third molar (M3) as percentages			
Study	GPF opposite M2	GPF between M2 and M3	GPF opposite M3	GPF posterior to M3
(Tomaszewska et al. 2014)	16.3	6.8	74.7	2.2
(Aoun et al. 2015)	1.7	27.6	41.4	29.3
(Sarilita and Soames 2016)	4.0	37.3	58.7	0.0
(Cagimni et al. 2017)	2.1	34.6	41.7	21.7

inferior-anterior-lateral appearing in 5.3 and 4.0% of cases, respectively (Tomaszewska et al. 2015).

The GPC also varies significantly in its path through the palatine bone. In the sagittal plane, it most often travels inferiorly at first and then slopes anterior-inferior part way through the canal, though the second most common path is a continuous anterior-inferior canal (Tomaszewska et al. 2015). In the coronal plane, most GPCs studied travel in an inferior-lateral direction at first and then turn either directly inferior or inferior-medial, with equal frequency (Tomaszewska et al. 2015). Less commonly, it travels directly inferior throughout its length. There are only rare cases of the GPC traveling in directions different from those described above.

The greater palatine artery and nerve also have varying anatomy. The nerve usually splits into two branches but has been noted to split into as many as four, the division into trunks or branches usually happening within the GPC (Hafeez et al. 2015). The artery, in contrast, is highly branched along its course over the palate, most branches coming off it toward the alveolar side of the palate at the levels of the second and third molars and the first and second premolars (Klosek and Rungruang 2009). The greater palatine neurovascular bundles can be found 7–17 mm from the cementoenamel junctions of the upper molars and premolars (Reiser et al. 1996), although this distance varies with bone loss.

10.3 Measurements and Morphometrics of the GPF and GPC

In addition to the shape and direction of the GPF and GPC, their lengths, diameters, and distances from local landmarks are of practical importance for surgeons. Most recent studies show the GPC to be 31.1 ± 2.9 mm long, consistent with earlier studies showing measurements of 29 ± 3 mm (Howard-Swirzinski et al. 2010; Tomaszewska et al. 2015). The diameter of the GPF has been measured in both the anteroposterior and mediolateral directions; the average in both cases is 2.7 ± 0.5 mm when measured in dry skulls (Cagimni et al. 2017). The angle of entry of the GPF through the palate is approximately 90 degrees (Gibelli et al. 2017).

The GPF has also been studied in relation to the anatomical landmarks of the palate, namely, the median palatine suture, the posterior palatal border, and the alveolar ridge. The mean distances in dry skulls are 16.16 ± 1.61 mm from the median palatine suture, 4.4 ± 1.27 mm from the posterior palatal border, and 4.06 ± 0.3 mm from the alveolar ridge (Cagimni et al. 2017). Other measurements include the distance between the medial walls of the GPF, found to be 27.6 ± 2.8 mm (Sarilita and Soames 2016). Taking measurements can be more complicated in the clinical setting when soft tissue is considered. To address this, recent studies have attempted needle insertions in cadavers. One such measurement was the distance from the insertion of a needle aimed at the GPF into the foveola palatina in the anteroposterior plane. The distance ranged from 1.1 to 8.3 mm (Iwanaga et al. 2017).

10.4 Laterality and Sex Differences

Since there are two GPFs in most patients, it is important from a clinical perspective to consider differences between sexes as well as between the individual patient's left and right sides.

Computed tomography studies on the palate have shown significantly larger bone structures across all measurements in males (Tomaszewska et al. 2014). Correlating with this effect of sex on bone structure, males have such significantly longer GPCs that the length can be used to judge the sex of the skull with 78% reliability (Tomaszewska et al. 2014). One study showed that females generally have a more anterior GPF than males (Gibelli et al. 2017). Shape, angle of entry of the GPC through the palate, and other anatomical variations of the GPF and GPC do not appear to be influenced by sex. The confounding factors of environment and genetics make it difficult to assess and determine a correlation (Tomaszewska et al. 2015; Gibelli et al. 2017).

Studies of dry skulls have revealed no significant differences between the left and right positions of the GPF relative to the molars, the diameters of the GPF, or the distances from surrounding landmarks including the median palatine suture, the posterior palatal border, and the alveolar ridge (Cagimni et al. 2017). However, some asymmetry was noted in the number of GPFs per side; there were cases of triple foramina on one side and a single foramen on the opposite side (Cagimni et al. 2017). Computed tomography studies, specifically on a Lebanese population, showed signs of asymmetry in the location of the GPF relative to the molars (Aoun et al. 2015). Although there are no recurrent patterns of asymmetry across studies, asymmetry of the GPF should be considered very likely in any patient as GPF development is linked to the development of the facial bones, which depends on many genetic, environmental, and functional activity factors (Rossi et al. 2003; Tomaszewska et al. 2015). Although the anatomy can differ between sides, no statistically significant side difference has been seen in GPC lengths or in angles of GPF entry through the palate (Sheikhi et al. 2013; Gibelli et al. 2017). Differences between men and women in the branching of the greater palatine artery have been noted (Fig. 10.7). Males are

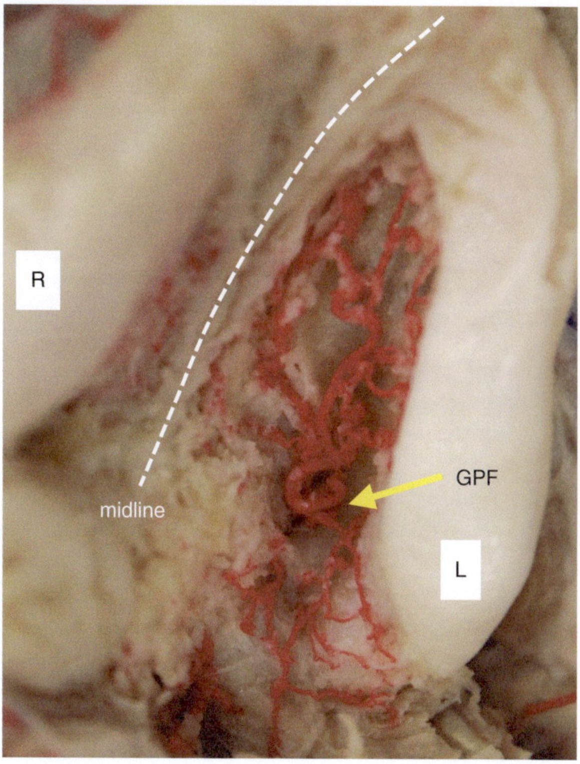

Fig. 10.7 Greater palatine artery (red) arising from greater palatine foramen (GPF)

more likely to have most the branching at the level of the first and second premolars, while females most often branch at the first and second molars (Klosek and Rungruang 2009).

10.5 Clinical Considerations

The GPF can serve as the landmark at which to aim in a search for the descending palatine artery and greater palatine nerve, but it also acts as a portal through which the pterygopalatine fossa can be reached (Hwang et al. 2011). It is therefore an important structure for dentists, periodontal surgeons, maxillofacial surgeons, ENT specialists, and neurologists. In particular, it can be useful in palatal surgery, restoration or extraction of the maxillary teeth, and endoscopic sinus surgery. Other clinical considerations outside the oral cavity include the use of block anesthesia for nasal intubation (Baddour et al. 1979). The GPF has been proposed to form part of a right triangle consisting of the palate, the posterior nasal spine, and the incisive foramina, which has proved consistent through skulls regardless of sex or laterality

(Gibelli et al. 2017). This consistency in structure across skulls makes it a reliable point of access, but as with most structures, anatomical variations can still complicate any procedure.

One of the most common procedures cited when the GPF and the GPC are considered is maxillary nerve block. Since the only absolute contraindication is swelling in the area of the GPF, this block is particularly useful (Wong and Sved 1991). A GPC that is longer than expected can entail inadequate infiltration of the pterygopalatine fossa, leading to inadequate anesthesia (Aoun et al. 2015). Further advancement of an anesthetic needle should be considered for the safety and comfort of the male patient. The anatomy should be thoroughly considered so as to avoid complications including neural injury, direct injection into a vessel, and excessive bleeding (Sved et al. 1992).

During maxillofacial surgeries such as septorhinoplasty, endoscopic sinus surgery, or posterior epistaxis, the GPF can also serve as a point of hemostasis (Williams and Ghorayeb 1990; Douglas and Wormald 2006). Again, adequate depth for the length of the GPC should be considered. Furthermore, the possibility of multiple GPFs should be considered during such procedures when hemostasis is difficult to establish, and it should be remembered that the number of GPFs can differ between left and right within the same patient (Cagimni et al. 2017).

The most common method for anesthesia via the GPF is to use a needle bent at the hub to an angle of 30–45 degrees and advancing the tip about 30 mm (Mercuri 1979 and Bassett K et al. 2015). For hemostasis, specifically during sinus surgery, the needle only needs to infiltrate as deep as 25 mm (Douglas and Wormald 2006).

Furthermore, the anatomy of the neurovasculature traveling through the GPF is important for harvesting tissue for free gingival grafts to treat diseases such as gingival recession (Jenabian et al. 2016). As the material is harvested from the alveolar side of the palate, the anatomy of the GPC, specifically the greater palatine artery, must be considered as most of the arterial branches to the palate extend in that direction (Klosek and Rungruang 2009; Jenabian et al. 2016). With the noted differences in branching between males and females, exploring a more posterior site can be considered in males.

Access to the pterygopalatine ganglion through the GPF has also been considered when new neuromodulation and nerve stimulators are explored for controlling migraine and cluster headaches and for nerve stimulation after a stroke (Oluigbo et al. 2011).

In summary, the most common location of the GPF is opposite the third molar. The position varies but it is rarely as anterior as the second molar. In patients lacking dentition, estimates from the median palatine suture, the posterior palatine border, and the alveolar ridge can serve as guides to the position of the GPF. Access to the pterygopalatine fossa for anesthesia and hemostasis should properly be performed with a needle bent at the hub to 30 degrees and advanced to 30 mm, though careful consideration should be given to males with longer GPCs and to patients who need bilateral infiltration, as asymmetry is not uncommon.

Lastly, surgeons should be aware of variants such as palatine bridges over the GPF or canals (Hauser and De Stefano 1989) (Fig. 10.8) in order to locate the GPF correctly (Iwanaga et al. 2017).

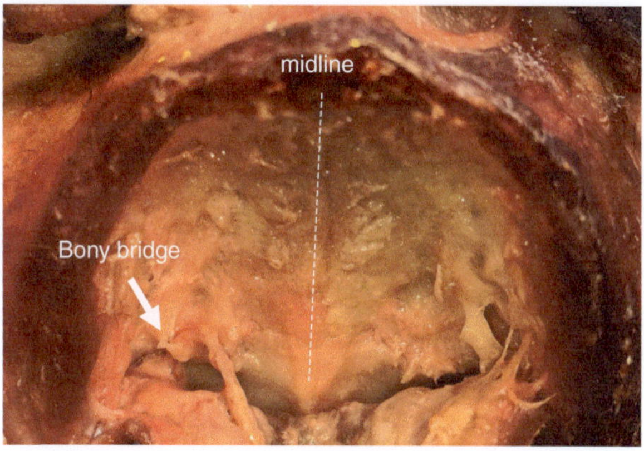

Fig. 10.8 Palatine bridges over the greater palatine foramen

References

Aoun G, Nasseh I, Sokhn S, Saadeh M (2015) Analysis of the greater palatine foramen in a Lebanese population using cone-beam computed tomography technology. J Int Soc Prev Community Dent 5:S82–S88. https://doi.org/10.4103/2231-0762.171594

Baddour HM, Hubbard AM, Tilson HB (1979) Maxillary nerve block used prior to awake nasal intubation. Anesth Prog 26:43–45

Bassett K, DiMarco A, Naughton D (2015) Local anesthesia for dental professionals. 2nd ed. Pearson, London

Cagimni P, Govsa F, Ozer MA, Kazak Z (2017) Computerized analysis of the greater palatine foramen to gain the palatine neurovascular bundle during palatal surgery. Surg Radiol Anat 39:177–184. https://doi.org/10.1007/s00276-016-1691-0

Douglas R, Wormald P-J (2006) Pterygopalatine fossa infiltration through the greater palatine foramen: where to bend the needle. Laryngoscope 116:1255–1257. https://doi.org/10.1097/01.mlg.0000226005.43817.a2

Gibelli D, Borlando A, Dolci C et al (2017) Anatomical characteristics of greater palatine foramen: a novel point of view. Surg Radiol Anat 39:1359–1368. https://doi.org/10.1007/s00276-017-1899-7

Hafeez NS, Ganapathy S, Sondekoppam R et al (2015) Anatomical variations of the greater palatine nerve in the greater palatine canal. J Can Dent Assoc 81:f14

Hauser G, De Stefano GF (1989) Epigenetic variants of the human skull. Schweizerbart, Stuttgart

Howard-Swirzinski K, Edwards PC, Saini TS, Norton NS (2010) Length and geometric patterns of the greater palatine canal observed in cone beam computed tomography. Int J Dent 2010:1. https://doi.org/10.1155/2010/292753

Hwang SH, Seo JH, Joo YH et al (2011) An anatomic study using three-dimensional reconstruction for pterygopalatine fossa infiltration via the greater palatine canal. Clin Anat 24:576–582. https://doi.org/10.1002/ca.21134

Iwanaga J, Voin V, Nasseh AA et al (2017) New supplemental landmark for the greater palatine foramen as found deep to soft tissue: application for the greater palatine nerve block. Surg Radiol Anat 39:981–984. https://doi.org/10.1007/s00276-017-1829-8

Jenabian N, Bahabadi MY, Bijani A, Rad MR (2016) Gingival unit graft versus free gingival graft for treatment of gingival recession: a randomized controlled clinical trial. J Dent (Tehran) 13:184–192

Klosek SK, Rungruang T (2009) Anatomical study of the greater palatine artery and related structures of the palatal vault: considerations for palate as the subepithelial connective tissue graft donor site. Surg Radiol Anat 31:245–250. https://doi.org/10.1007/s00276-008-0432-4

Mercuri LG (1979) Intraoral second division nerve block. Oral Surg Oral Med Oral Pathol 47:109–113

Oluigbo CO, Makonnen G, Narouze S, Rezai AR (2011) Sphenopalatine ganglion interventions: technical aspects and application. In: Peripheral nerve stimulation. KARGER, Basel, pp 171–179

Reiser GM, Bruno JF, Mahan PE, Larkin LH (1996) The subepithelial connective tissue graft palatal donor site: anatomic considerations for surgeons. Int J Periodontics Restorative Dent 16(2):130–137

Rossi M, Ribeiro E, Smith R (2003) Craniofacial asymmetry in development: an anatomical study. Angle Orthod 73:381–385. https://doi.org/10.1043/0003-3219(2003)073<0381:CAIDAA>2.0.CO;2

Sarilita E, Soames R (2016) Morphologyof the hard palate: a study of dryskulls and review of the literature. Morfología del paladar duro: un studio en cráneos secos y revision de la literatura. Rev Argentina Anatomía Clínica 7:34–43

Sheikhi M, Zamaninaser A, Jalalian F (2013) Length and anatomic routes of the greater palatine canal as observed by cone beam computed tomography. Dent Res J (Isfahan) 10:155–161

Sved AM, Wong JD, Donkor P et al (1992) Complications associated with maxillary nerve block anaesthesia via the greater palatine canal. Aust Dent J 37:340–345

Tomaszewska IM, Frączek P, Gomulska M et al (2014) Sex determination based on the analysis of a contemporary polish population's palatine bones: a computed tomography study of 1,200 patients. Folia Morphol (Warsz) 73:462–468. https://doi.org/10.5603/FM.2014.0069

Tomaszewska IM, Kmiotek EK, Pena IZ et al (2015) Computed tomography morphometric analysis of the greater palatine canal: a study of 1,500 head CT scans and a systematic review of literature. Anat Sci Int 90:287–297. https://doi.org/10.1007/s12565-014-0263-9

Tomaszewska IM, Tomaszewski KA, Kmiotek EK et al (2014) Anatomical landmarks for the localization of the greater palatine foramen--a study of 1200 head CTs, 150 dry skulls, systematic review of literature and meta-analysis. J Anat 225:419–435. https://doi.org/10.1111/joa.12221

Williams WT, Ghorayeb BY (1990) Incisive canal and pterygopalatine fossa injection for hemostasis in septorhinoplasty. Laryngoscope 100:1245–1247. https://doi.org/10.1288/00005537-199011000-00022

Wong JD, Sved AM (1991) Maxillary nerve block anaesthesia via the greater palatine canal: a modified technique and case reports. Aust Dent J 36:15–21

Anatomy and Variations of the Incisive Foramen

11

Iwona M. Tomaszewska, Patrick Popieluszko,
Krzysztof A. Tomaszewski, and Jerzy A. Walocha

11.1 Introduction

The incisive foramen (Fig. 11.1) is the distal opening of the incisive canal and an important structure to consider during dental implant, cystectomy of radicular and nasopalatine cysts, and extraction of supernumerary mesiodens (Naitoh et al. 2015). It serves as the exit point for the nasopalatine nerve and sphenopalatine artery (Fig. 11.2). Being aware of the structure and its possible variations helps surgeons operating in the area to avoid injuring the neurovascular bundle and to direct local anesthesia appropriately. There have been several studies of the variations and their implications (Asaumi et al. 2010; Bornstein et al. 2011; Tözüm et al. 2012).

Embryologically, the fusing lateral palatal shelves overlap with the anterior primary palate, as indicated later by the sloping pathways of the junctional incisive neurovascular canals that carry the previously formed incisive nerves and blood vessels. The palate begins to ossify during the 8th week postconception (Sperber et al. 2010).

In this chapter, the normal anatomy, the possible variations, and the clinical importance of the incisive foramen and canal are discussed.

11.2 Incisive Foramen, Incisive Canal, Nasopalatine Nerve, and Sphenopalatine Artery

The incisive foramen (IF) lies in the bony palate, directly behind the first two incisors, in the incisive fossa (Fig. 11.3). It carries the sphenopalatine artery and the nasopalatine nerve from the nasal cavity through the bony palate via the incisive

I. M. Tomaszewska (✉) · P. Popieluszko · K. A. Tomaszewski · J. A. Walocha
Department of Anatomy, Jagiellonian University Medical College, Krakow, Poland

J. Iwanaga, R. S. Tubbs (eds.), *Anatomical Variations in Clinical Dentistry*,
https://doi.org/10.1007/978-3-319-97961-8_11

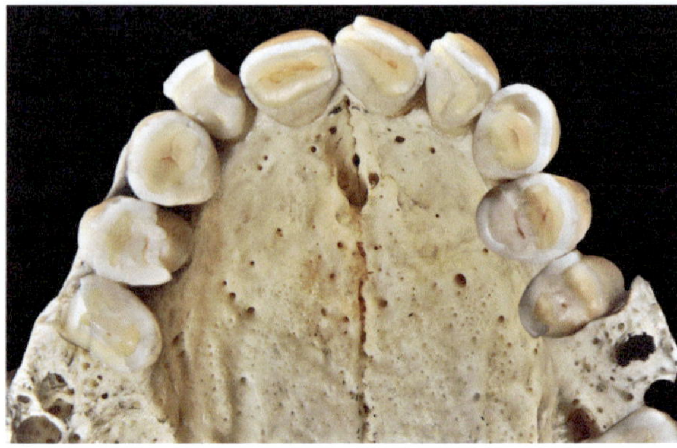

Fig. 11.1 Bony specimen (hard palate) showing the incisive foramen

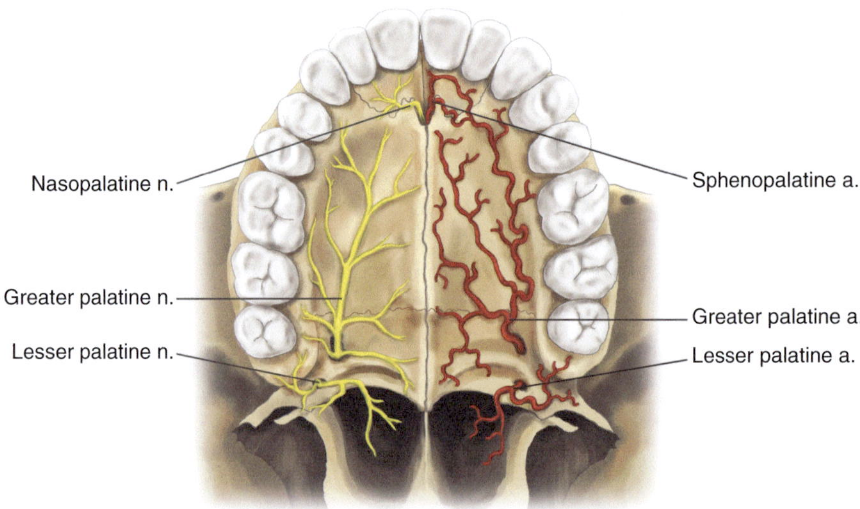

Fig. 11.2 The incisive foramen and greater and lesser palatine foramina, with vessels and nerves exiting them (drawing by Karolina Saganiak MD)

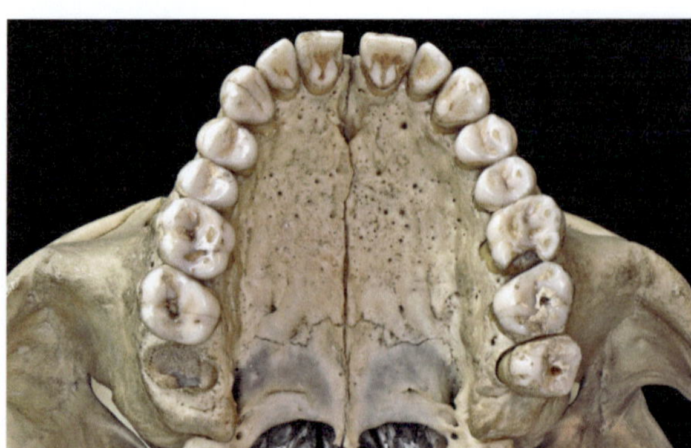

Fig. 11.3 Bony specimen (hard palate) showing the relationship of the incisive foramen to the incisors, the median palatine suture, and the greater and lesser palatine foramina

canal (IC) (Fig. 11.4). Depending on the morphology of the IC, there can be one or multiple foramina. Three-dimensional micro-CT analysis demonstrates a "Y-morphology" of the canal in about 60% of the population, which leads to two or multiple foramina in the palate, while other variations can lead to a single foramen in this place (Fukuda et al. 2015). The canal itself has a characteristic shape when observed from the sagittal plane, typically cylinder, banana, funnel, and hourglass shapes, the funnel shape being the most common (Fukuda et al. 2015) (Fig. 11.5).

A recent study of Lebanese patients showed the IF to have an anterior-posterior diameter of 4.91 mm with an average entry angle of 17.09 degrees and an average length of 11.52 mm of the IC (Nasseh et al. 2017). These measurements were comparable to those found among other nationalities (Table 11.1).

Naitoh et al. (2015) measured the anteroposterior axis of the IF between Hellman's dental age IA and IIIC and found values mostly in the range 0.78–1.48 mm.

The nasopalatine nerve and the fibers provide sensory innervation to the mucosa of the hard palate around all four incisors and the canines. The sphenopalatine artery enters the oral cavity through the IC, supplies the mucosa of the hard palate around the incisors and canines, and anastomoses with the greater palatine artery. Its anatomy can also vary: a bony ridge can separate the foramen through which it enters

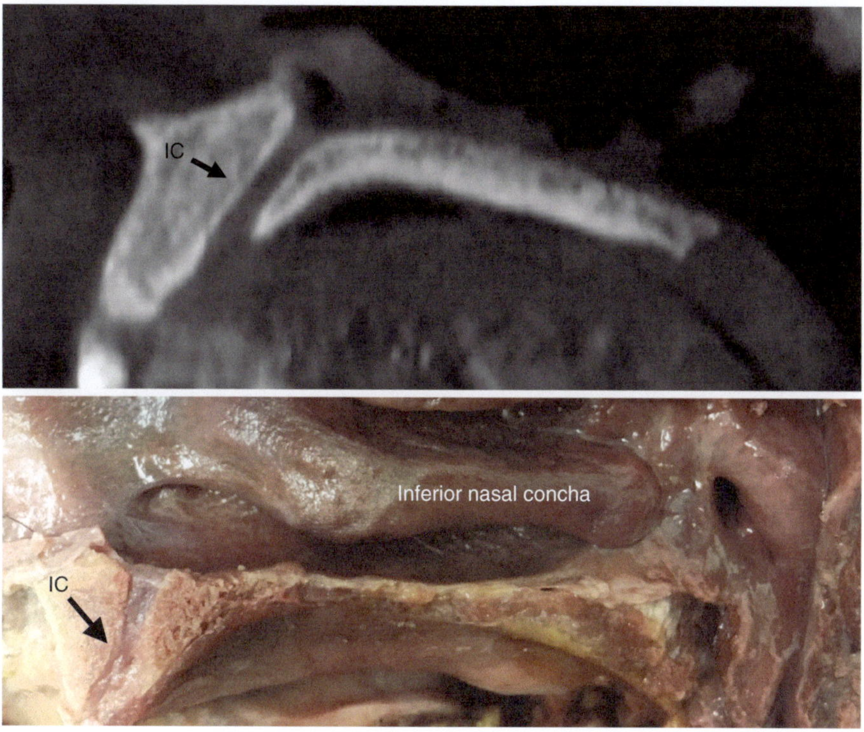

Fig. 11.4 Incisive canal (IC) in sagittal plane

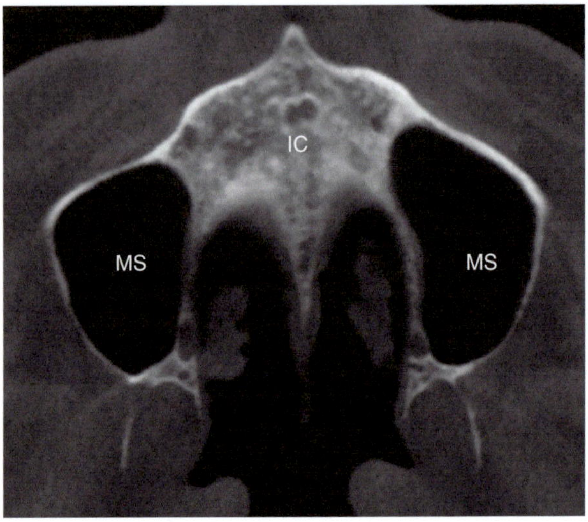

Fig. 11.5 Incisive canal (IC) in axial plane. MS; maxillary sinus

Table 11.1 Morphometrics of the IF and IC across studies

Study	Diameter of IF (mm)	Length of IC (mm)
Mraiwa et al. (2004)	4.3 ± 0.9	8.1 ± 3.4
Salemi et al. (2016)	4.74 ± 1.11	–
Al-Amery et al. (2015)	2.8 ± 0.82	16.33 ± 4.43
Kajan et al. (2015)	3.5 ± 1.1	12.84 ± 2.88
Thakur et al. (2013)	–	10.08 ± 2.25
Tözüm et al. (2012)	–	10.86 ± 2.67
Fukuda et al. (2015)	–	11.75 ± 1.70

Data presented as mean ± standard deviation

the nasal cavity, it can have a superior ramification point, and its number of branches can vary (Midilli et al. 2015). The most common branching pattern is two branches superiorly and one inferiorly, but this is found in only 25% of cases; on average there are 2.4 branches superiorly and 2.1 inferiorly (Eordogh et al. 2017).

11.3 Special Considerations for Cleft Lip and Palate

The IF and IC can be included in cleft palate deformities; however, the IF has been used specifically to grade the extent of submucosal cleft palates. A submucosal cleft palate including a deformity of the IC or IF is considered severe (Mori et al. 2013). The Kernahan "Y" classification system for cleft lips and palates uses the incisive foramen as a landmark, corresponding to the point where the three lines of the "Y" join, and numbers along the lines to denote the zone in which the cleft is located. This system has been modified many times. The revised Smith-modified Kernahan "Y" classification denotes type 7 as a separate line in the expanded "Y" corresponding to a submucosal cleft palate, alphabetical labeling along the line denoting the extent of the cleft, 7a being a cleft reaching anterior to the IF (Khan et al. 2013).

11.4 Clinical Considerations

The IF and its contents are of the utmost importance in placing an implant for a lost maxillary incisor. Bone material for an implant can be scarce and a large IF can compromise what little there is (Artzi et al. 2000). A CT study of the position of the IF in relation to planned incisor implants showed that the anatomy would compromise the neurovasculature of the anterior hard palate in 4% of the patients studied (Kraut and Boyden 1998). The most recent meta-analysis shows that for implant placement in this area, the success rate is 84–100%, most common complication being temporary sensory loss in the anterior palate (de Mello et al. 2017). Caution is urged as cases have been noted in which patients reported no resolution of sensory loss after implant placement follow-up, suggesting permanent injury to the nasopalatine nerves (de Mello et al. 2017). Artzi et al. (2000) described a case of a maxilla with a large IF that had to be augmented with a bone graft to push back the neurovasculature contents of the IF in order to preserve the blood supply and specifically the sensation of the hard palate. They noted the importance of maintaining phonetic and functional outcomes in the young male, which would have been compromised without proper planning and injury to the nasopalatine nerve (Artzi et al. 2000).

The anatomy of the IF and the IC differs between the sexes. Men have consistently larger IF and IC diameters (Tözüm et al. 2012; Thakur et al. 2013; Al-Amery et al. 2015; Kajan et al. 2015; Nasseh et al. 2017). Other studies have shown that age does not affect the dimensions of the IF or the IC but contributes to the loss of maxillary bone thickness. This demonstrates the particular relevance of IF and IC anatomy to older male patients considering incisor implants.

The nasopalatine nerve can be targeted for local anesthesia for procedures involving the incisors, e.g., extractions of the maxillary central supernumery teeth. Injecting in the submucosa is considered painful, but attempts to find a more convenient and less painful site for injection such as labial infiltration have shown total anesthetic efficacy of only 76.7% while the traditional method provides 83% (Lassemi et al. 2008). To date, there has been no research on the effect of IF variations on the efficacy of injecting at this site.

An artery considered for embolization for posterior epistaxis is the descending palatine artery, the origin of the greater palatine artery, which returns to the posterior nares through the IF (Krajina and Chrobok 2014). Because the anatomy of the sphenopalatine artery is variable, especially in its numerous possible branches, hemostasis through the descending palatine artery can be considered. The patency of the IF in patients with chronic epistaxis could serve as a source of bleeding but could also serve as an easily accessible point for compression of the artery to help stop such bleeding. Radiographic studies and cases of patients treated with such cautery of the greater palatine artery have confirmed the viability of this technique for anterior epistaxis, the cases described showing resolution of the epistaxis up to 24 months after the procedure (Butrymowicz et al. 2016).

References

Al-Amery SM, Nambiar P, Jamaludin M et al (2015) Cone beam computed tomography assessment of the maxillary incisive canal and foramen: considerations of anatomical variations when placing immediate implants. PLoS One 10:e0117251. https://doi.org/10.1371/journal.pone.0117251

Artzi Z, Nemcovsky CE, Bitlitum I et al (2000) Displacement of the incisive foramen in conjunction with implant placement in the anterior maxilla without jeopardizing vitality of nasopalatine nerve and vessels: a novel surgical approach. Clin Oral Implants Res 11:505–510. https://doi.org/10.1034/j.1600-0501.2000.011005505.x

Asaumi R, Kawai T, Sato I et al (2010) Three-dimensional observations of the incisive canal and the surrounding bone using cone-beam computed tomography. Oral Radiol 26:20–28

Bornstein MM, Balsiger R, Sendi P et al (2011) Morphology of the nasopalatine canal and dental implant surgery: a radiographic analysis of 100 consecutive patients using limited cone-beam computed tomography. Clin Oral Implants Res 22:295–301

Butrymowicz A, Weisstuch A, Zhao A et al (2016) Endoscopic endonasal greater palatine artery cauterization at the incisive foramen for control of anterior epistaxis. Laryngoscope 126:1033–1038. https://doi.org/10.1002/lary.25677

de Mello JS, Faot F, Correa G et al (2017) Success rate and complications associated with dental implants in the incisive canal region: a systematic review. Int J Oral Maxillofac Surg 46:1584–1591. https://doi.org/10.1016/j.ijom.2017.05.002

Eordogh M, Grimm A, Gawish I et al (2018) Anatomy of the sphenopalatine artery and its implications for transnasal neurosurgery. Rhinology 56:82–88. https://doi.org/10.4193/Rhin17.181

Fukuda M, Matsunaga S, Odaka K et al (2015) Three-dimensional analysis of incisive canals in human dentulous and edentulous maxillary bones. Int J Implant Dent 1:12. https://doi.org/10.1186/s40729-015-0012-4

Kajan ZD, Kia J, Motevasseli S et al (2015) Evaluation of the nasopalatine canal with cone-beam computed tomography in an Iranian population. Dent Res J (Isfahan) 12:14–19

Khan M, Ullah H, Naz S et al (2013) A revised classification of the cleft lip and palate. Can J Plast Surg 21:48–50

Krajina A, Chrobok V (2014) Radiological diagnosis and management of epistaxis. Cardiovasc Intervent Radiol 37:26–36. https://doi.org/10.1007/s00270-013-0776-y

Kraut RA, Boyden DK (1998) Location of incisive canal in relation to central incisor implants. Implant Dent 7:221–225

Lassemi E, Motamedi MHK, Jafari SM et al (2008) Anaesthetic efficacy of a labial infiltration method on the nasopalatine nerve. BDJ 205:E21–E21. https://doi.org/10.1038/sj.bdj.2008.872

Midilli R, Gode S, Ozturk K (2015) The "two-point" technique for endoscopic sphenopalatine artery cauterization: is it effective and safe? Hippokratia 19:284

Mori Y, Hoshi K, Takato T et al (2013) Submucous cleft palate: variations in bony defects of the hard palate. Br J Oral Maxillofac Surg 51:e220–e223. https://doi.org/10.1016/j.bjoms.2013.01.015

Mraiwa N, Jacobs R, Van Cleynenbreugel J et al (2004) The nasopalatine canal revisited using 2D and 3D CT imaging. Dentomaxillofacial Radiol 33:396–402. https://doi.org/10.1259/dmfr/53801969

Naitoh M, Arikawa T, Nishiyama W et al (2015) Observation of maxillary incisive canal using dry skulls between Hellman's dental age IA and IIIC. Okajimas Folia Anat Jpn 92:37–42. https://doi.org/10.2535/ofaj.92.37

Nasseh I, Aoun G, Sokhn S (2017) Assessment of the Nasopalatine Canal: an anatomical study. Acta Inform Med 25:34–38. https://doi.org/10.5455/aim.2017.25.34-38

Salemi F, Moghadam FA et al (2016) Three-dimensional assessment of the nasopalatine canal and the surrounding bone using cone-beam computed tomography. J Periodontal Implant Dent 8:1–7. https://doi.org/10.15171/jpid.2016.001

Sperber GH, Sperber SM, Guttmann GD (2010) Palate. In: Craniofacial embryogenetics and development, 2nd edn. People's Medical Publishing House-USA, Shelton, CT, pp 131–144

Thakur AR, Burde K, Guttal K et al (2013) Anatomy and morphology of the nasopalatine canal using cone-beam computed tomography. Imaging Sci Dent 43:273–281. https://doi.org/10.5624/isd.2013.43.4.273

Tözüm TF, Güncü GN, Yıldırım YD et al (2012) Evaluation of maxillary incisive canal characteristics related to dental implant treatment with computerized tomography: a clinical multicenter study. J Periodontol 83:337–343. https://doi.org/10.1902/jop.2011.110326

Variant Anatomy of the Torus Palatinus 12

Tatsuo Okui

12.1 Introduction

The torus palatinus (TP), a benign osseous outgrowth of the oral bony palate, is a bone mass that accrues along the midline of the hard palate. It represents an anatomical variation rather than pathology. TP generally causes no symptoms and needs no treatment. Surgical removal is required in cases of chronic trauma or if there is interference with oral function or denture replacement. TPs have recently been used as autogenous bone graft material for alveolar ridge reconstruction during dental implant treatments (Moraes Junior et al. 2010). They have been reported in different regions and are associated with age, gender, diet, and other environmental factors.

12.2 Classification of TPs

TPs are classified on the basis of their location, size, and shape.

12.2.1 Location

Most TPs are located in the molar-premolar area (Table 12.1). Gorsky (1996) reported that in a younger age group, the majority were located in the molar area. King and Moore 1971 described an age-related change of prevalence of TP from the molar to the molar-premolar area.

T. Okui (✉)
Department of Oral and Maxillofacial Surgery, Okayama University Graduate School of Medicine, Dentistry and Pharmaceutical Sciences, Okayama, Japan
e-mail: pphz1rke@okayama-u.ac.jp

J. Iwanaga, R. S. Tubbs (eds.), *Anatomical Variations in Clinical Dentistry*,
https://doi.org/10.1007/978-3-319-97961-8_12

Table 12.1 Predilection site of TP (Gorsky et al. 1996; Sisman et al. 2008; Yildiz et al. 2005)

Year	Location			Study
	Molar (%)	Premolar (%)	Molar-Premolar (%)	
1996	53.8	10	36.2	Gorsky (Gorsky et al. 1996)
2005	62.9	4.4	32.5	Yildiz (Yildiz et al. 2005)
2008	15.4	13.6	66.4	Sisman (Sisman et al. 2008)

Table 12.2 Classification by size (Gorsky et al. 1996; Sathya et al. 2012; Sisman et al. 2008; Yildiz et al. 2005)

Year	Size		Population	Study
	>2 cm(%)	<2 cm(%)		
1996	31.9	68.1	Israel	Gorsky (Gorsky et al. 1996)
2005	8.5	91.5	Turkey school children (southern region of Turkey)	Yildiz (Yildiz et al. 2005)
2008	24.5	75.4	Turkey school children (cappadokia)	Sisman (Sisman et al. 2008)
2012	67.4	32.6	Malaysian	Sathya (Sathya et al. 2012)

12.2.2 Size

The prevalence of TPs <2 cm (68–91%) is much greater than that of larger ones. However, Sathya reported that Asian populations tend to have TPs >2 cm (Sathya et al. 2012) (Table 12.2).

12.2.3 Shape

Morphologically, TPs have assorted shapes: flat (smooth), spindle, lobular, and nodular (Fig. 12.1). The developmental processes resulting in the different shape types are not known. Several reports have revealed that the flat type of TP occurs in 58–63% of cases, the spindle shape in 23–55%, the nodular shape in 1–14%, and the lobular shape in 6–33% (Table 12.3).

12.3 Frequency of TPs

Many researchers have investigated the etiological factors involved in TP, but no consensus has been reached. The etiology of TP is currently thought to involve the interaction of multiple factors.

12.3.1 Race

Many investigators have proposed that racial or ethnic group differences affect the occurrence of TP. The prevalence of TP in various populations has revealed a

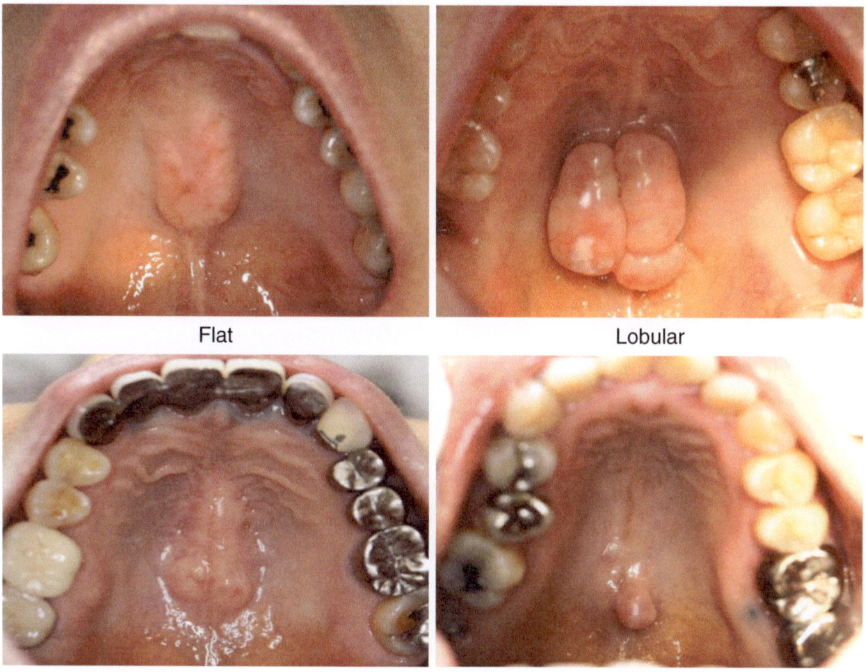

Fig. 12.1 Classification by shape of TP (Sisman et al. 2008; Hiremath et al. 2011; AlZarea 2016)

Table 12.3 Different type of TP

Year	Shape				Study
	Spindle(%)	Nodular(%)	Lobular(%)	Flat(%)	
2008	36.3	0.9	0	62.7	Sisman (Sisman et al. 2008)
2011	54.6	12.1	33.3	–	Hiremath (Hiremath et al. 2011)
2016	22.7	13.6	6.1	57.5	Alzarea (AlZarea 2016)

very wide range (1.4–66%) according to previous reports and a review. The highest prevalence (66%) was found in Asians and Eskimos; the lowest (1.4%) was reported in a study based in Saudi Arabia. It has been suggested that genetic differences could account for the higher prevalence of TP among populations in East Asia and adjacent parts of Central, Southeast, North, and South Asia (Table 12.4).

12.3.2 Age

The occurrence of TP generally increases with age, particularly by the third decade, a period when an individual's peak bone mass is usually achieved.

Table 12.4 Prevalence of TP in origin (Location) (Sisman et al. 2008; Yoshinaka et al. 2010; García-García et al. 2010; Al Quran and Al-Dwairi 2006; Jainkittivong et al. 2007)

Year	Population	Plevalence (%)	Study
1950	Eskimos	66	Woo
1950	Japanese	46.6	Woo
1953	United States	20.8	Kolas
1966	Yugoslavian	49.7	Vidic
1977	Brazilian Indian	10	Bernaba
1983	Malaysian	24.4	Hashim
1984	Singapore	48	Chew
1985	Icelendic, South-Thingeyjarsysla	33.3	Axelsson
1985	Icelendic, North-Thingeyjarsysla	14.6	Axelsson
1987	Saudi Arabia	1.4	Salem
1988	German	13.5	Reichart
1988	Thai	23.1	Reichart
1992	Indian	9.5	Shah
1992	Norway (Oslo)	9.22	Haugen
1994	Norway (Lofoten)	36.1	Eggen
1996	Vietnamese	6	Nair
1996	Israel	21	Gorsky
1999	Southern Thailand	61.7	Kerdpon
1999	Turkish	45.4	Gozil
1999	United States	22.8	Sonnier
2001	African	6.2	Al-bayaty
2001	West Indies	6.6	Al-bayaty
2002	Thai	58.1	Apinhasmit
2003	Black	36	Agnihotri
2003	Caucasian	17	Agnihotri
2004	Turkish	20.9	Cagirankaya
2005	Turkish	30.9	Yildiz (Yildiz et al. 2005)
2006	Jorden	29.8	Al Quran (Al Quran and Al-Dwairi 2006)
2007	Thai	60.5	Jainkittivong (Jainkittivong et al. 2007)
2007	Turky (cappadokia)	4.1	Sisman (Sisman et al. 2008)
2010	Japanese	17	Yoshinaka (Yoshinaka et al. 2010)

12.3.3 Geographical Location

The influence of geographical location on the prevalence of the TP has been shown to result from nutritional factors. Haugen (Haugen 1992) and Eggen et al. (1994) speculated that nutrients in saltwater fish (i.e., omega-3 polyunsaturated fatty acids and vitamin D, the most important osteogenesis factor) could account for the significantly higher prevalence of TP they observed in an island population (36.1% in Lofoten, Norway) than an inland population (9.2% in Oslo, Norway). A similar tendency was observed in Iceland by Axelsson and Hedegaard (1985).

In the Cappadocia region, the prevalence of TP is much lower than in other regions of Turkey (Sisman et al. 2008). Seafood consumption is not as common in the Cappadocia region as in the other regions studied, and this could again be implicated in the low prevalence. The TP prevalence in Japan in 1950 was approximately

50%, but in a 2010 study, it had decreased to 17% (Yoshinaka et al. 2010). Over that 60-year period, the dietary habits among Japanese changed from the traditional Japanese diet to a Westernized diet. The nutritional factors involved in this significant change could have influenced the prevalence of TP in Japan.

12.3.4 Sex

On the whole, women show a higher prevalence of TP than men (women:men = 1.5–2.0:1) (Table 12.5). This is likely to be attributable to a dominant X chromosome gene (Imada et al. 2014). This higher prevalence in women was observed in Asian populations but not among Indians, where the occurrence was almost equal between men and women. Belsky revealed that postmenopausal Caucasian women with large TPs have higher bone densities than their peers and higher bone densities than much younger women (Belsky et al. 2003).

12.4 Anatomy of Related Structures

12.4.1 Greater Palatine Foramen

A systematic review by Tomaszewska on the relationship between the greater palatine foramen (GPF) and the palatal structures revealed that the GPF was positioned 15.9 ± 1.5 mm from the median palatine suture, 3.0 ± 1.2 mm from the alveolar ridge, and 17.0 ± 1.5 mm from the posterior nasal spine; 74.7% of GFPs were positioned opposite the maxillary third molar (Tomaszewska et al. 2014).

Table 12.5 Prevalence of TP in sex (Sathya et al. 2012; Sisman et al. 2008; Yoshinaka et al. 2010; Yildiz et al. 2005; García-García et al. 2010; Jainkittivong et al. 2007)

Year	Famales (%)	Males (%)	Study
1984	48	48	Chew
1988	15.1	11.7	Reichart German population
1988	28.5	15.8	Reichart Thai population
1992	11.2	6.72	Haugen
1994	39.8	23.7	Eggen
1996	24.9	16.4	Gorsky
1999	67.6	48.1	Kerdpon
1999	25.9	14.8	Sonnier
2001	7.7	4.5	Al-bayaty
2002	67.3	48.8	Apinhasmit
2004	28.2	6	Cagirankaya
2005	34.3	28.1	Yildiz (Yildiz et al. 2005)
2006	47	14	Al Quran (Al Quran and Al-Dwairi 2006)
2007	70.5	48.8	Jainkittivong (Jainkittivong et al. 2007)
2007	5.7	1.8	Sisman (Sisman et al. 2008)
2010	24.6	7.5	Yoshinaka (Yoshinaka et al. 2010)
2012	15.7	8.4	Sathya (Sathya et al. 2012)

12.4.2 Greater Palatine Nerve and Artery

The greater palatine nerve is a bundle of nerves running from the GPF through the palatine canal, along the palatal groove, and communicating with the nasopalatine nerve. Miwa et al. (2018) reported the specific arrangement of the greater palatine nerve and arteries in the molar region. The greater palatine artery usually runs deep to the greater palatine nerve. The distribution of blood vessels and nerves becomes remarkable, and the number of the lateral branches increases at the molar and premolar areas of the palate.

12.4.3 The Thickness of the Bony Palate

Surgeons should be familiar with the normal thickness of the bony palate in order to avoid perforation into the nasal cavity. Kang et al. (2007) reported that the bony palate in the median palatine suture is ≥6 mm thick and that 6–10 mm laterally from the suture the normal bone thickness decreases to ≤3 mm (Fig. 12.2). In removing a wide flat-type TP, surgeons should carefully examine the TP volume and physiological bone thickness by coronal plane computed tomography (CT) images to preclude perforation of the nasal cavity.

12.5 Surgical Procedure for the Removal of a TP

A TP can be removed under local anesthesia (lidocaine 2% with epinephrine 1:80,000) or general anesthesia. Briefly, the surgical procedure begins with exposure of the TP, followed by a segmental osteotomy under irrigation, the removal of bone

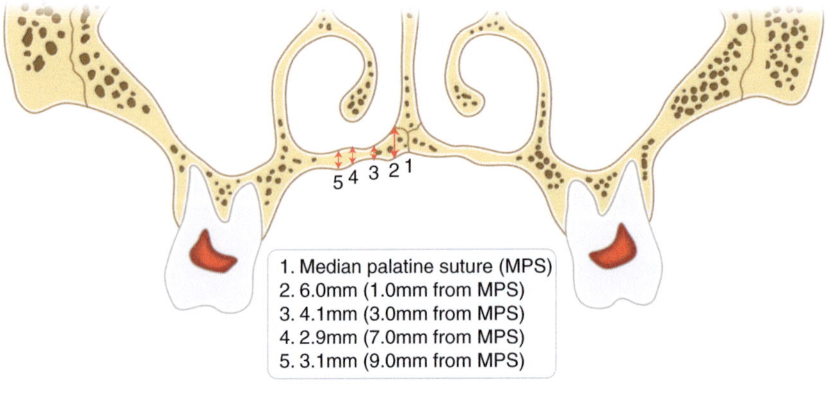

Fig. 12.2 Thickness of bony palate

fragments with a chisel, sutures, and compression. Pathological examination of the bone pieces is used to confirm the TP.

The removal is initiated by making two anterior and posterior oblique incisions (single-Y or double-Y incision) along the midline of the palate. This is designed to avoid injuring the greater palatine artery and nerve and provide access to the surgical field without excessive tension. A full-thickness flap is retracted with bilateral subperiosteal retractors. For a TP with deep grooves, a fine instrument should be used for periosteum peeling so the mucosa will not be torn. For a large TP, releasing the posterior part of the periosteum is easier after the bulk of the TP has been exposed and removed.

The procedure by which the TP is removed depends on its size. A large TP is sectioned with a fissure bur after complete exposure of the surgical field, and the segments are individually removed using an osteotome and mallet. The remaining sharp bony edge is smoothed with a bur or bone file. For small TPs, a large round bur can be used to smooth the bone. After the bone surface has been made smooth, excess soft tissue is trimmed and irrigated with saline, and the flaps are repositioned and sutured with interrupted sutures. Mattress or simple sutures should be used with less tension, and a surgical stent can be used to protect the wound during healing (Fig. 12.3).

When the TP is small, the initial incision is made in the midline. The procedure is then performed in exactly the way described above (Fig. 12.4).

12.5.1 Complications

Iatrogenic complications can occur as a result of manipulation by the surgeon, e.g., perforation to the nasal cavities, nerve and/or artery injuries, bone necrosis due to overheating with surgical drilling, hemorrhage because the branches of the greater palatine artery are injured, and fractures. Surgeons should therefore be properly prepared regarding the management of surgical approaches and possible complications.

12.5.2 Bisphosphonate-Related Osteonecrosis of the Jaw (BRONJ) on TP

Bisphosphonates (BPs) are frequently used to treat osteoporosis, bone metastases of malignant tumors (breast cancer, prostate cancer, and lung cancer), and multiple myeloma (MM) bone destruction. BPs induce the apoptosis of osteoclasts by inhibiting farnesyl diphosphate synthase, leading to suppression of bone resorption and bone remodeling.

In 2003, Marx reported the first case of osteonecrosis of the jaw in a patient with cancer and osteoporosis who had been treated with BPs (Marx 2003). Subsequently, many investigators have reported similar cases of osteonecrosis of the jaw following treatment with BPs; the condition was eventually named bisphosphonate-related osteonecrosis of the jaw (BRONJ). The inhibition of bone remodeling by BPs

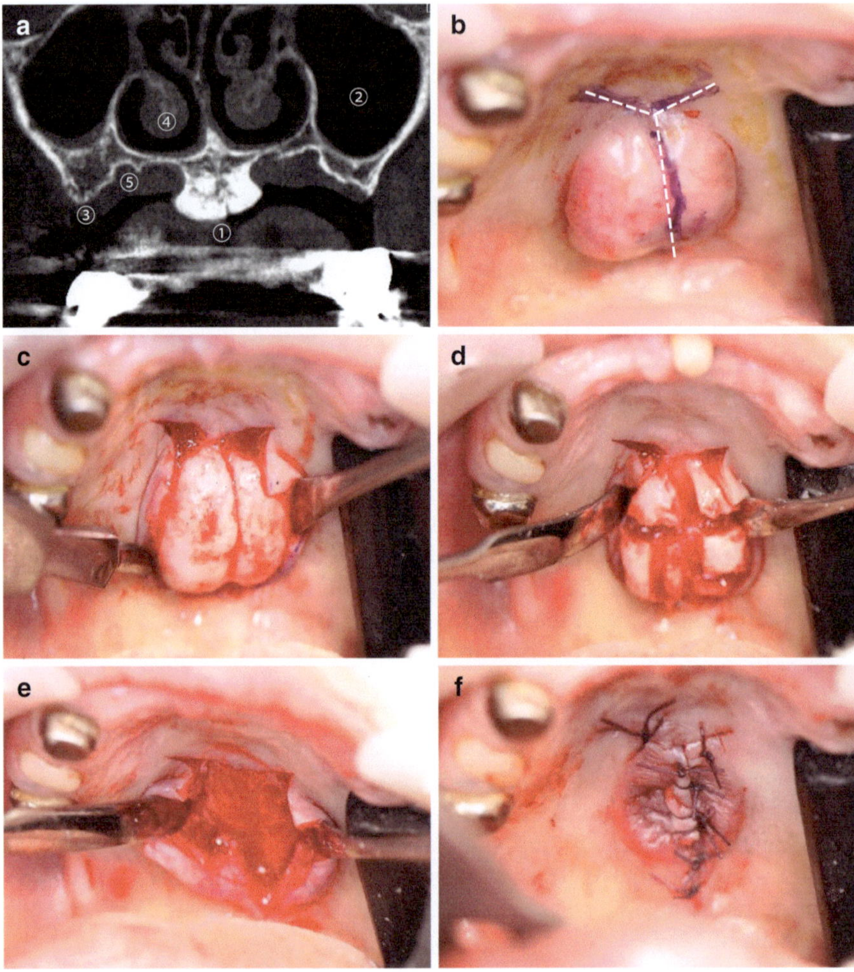

Fig. 12.3 Surgical procedure for removal large TP. (**a**) Coronal CT image of TP: 1. TP; 2. Maxillary sinus; 3.Alveolar bone; 4. Nasal cavity; 5. Palatine grooves. (**b**) Single Y incision. (**c**) Periosteum peeling. (**d**) Segmental osteotomy. (**e**) View of the hard palate after the surgical removal of the TP. (**f**) Suture

interferes with healing and increases the likelihood of infection. BRONJ frequently occurs in the mandible or maxilla in association with tooth extraction or periodontal surgery. However, Godinho and Kaneko reported cases of BRONJ in individuals with TPs (Godinho et al. 2013; Kaneko and Takahashi 2014). Large TPs, especially the lobular type, are covered by a thin and poorly vascularized mucosa. In these cases, healing after a traumatic injury will be inhibited and a bacterial infection can develop. This could be a risk factor of BRONJ in TP cases. There is no consensus regarding the treatment of BRONJ in individuals with a TP. However, discontinuation of BP administration and surgical removal of the TP should be considered in cases of intractable pain associated with BRONJ.

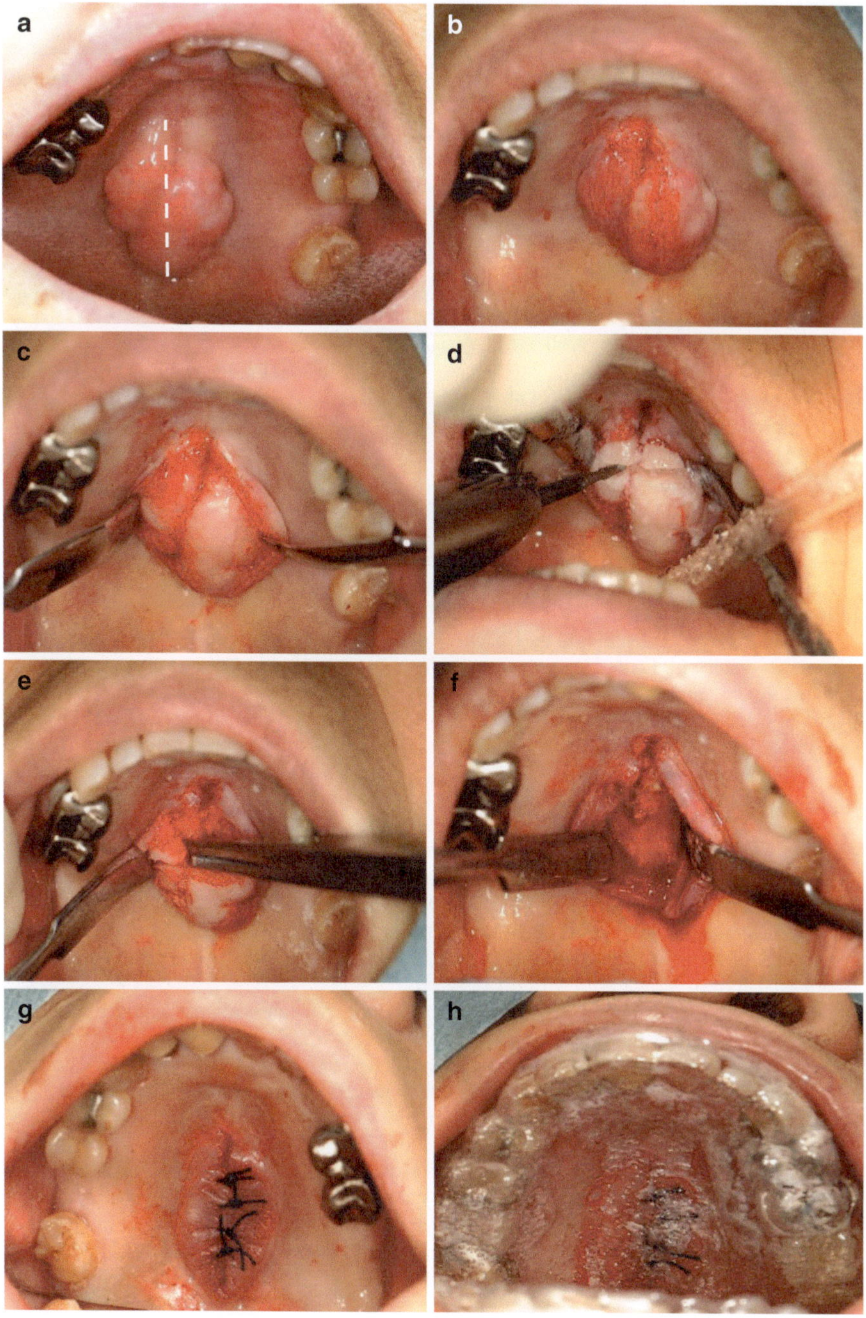

Fig. 12.4 Surgical procedure for removing small TP. (**a**) Straight incision line. (**b**) After straight incision. (**c**) Periosteum peeling. (**d**) Segmental osteotomy. (**e**) Removal of bone pieces with osteotome. (**f**) View of the hard palate after surgical removal of the TP. (**g**) Suture. (**h**) Surgical protector placement

References

Al Quran FAM, Al-Dwairi ZN (2006) Torus palatinus and torus mandibularis in edentulous patients. J Contemp Dent Pract 7:112–119

AlZarea BK (2016) Prevalence and pattern of torus palatinus and torus mandibularis among edentulous patients of Saudi Arabia. Clin Interv Aging 11:209–213

Axelsson G, Hedegaard B (1985) Torus palatinus in Icelandic schoolchildren. Am J Phys Anthropol 67:105–112

Belsky JL, Hamer JS, Hubert JE et al (2003) Torus palatinus: a new anatomical correlation with bone density in postmenopausal women. J Clin Endocrinol Metab 88:2081–2086

Eggen S, Natvig B, Gåsemyr J (1994) Variation in torus palatinus prevalence in Norway. Scand J Dent Res 102:54–59

García-García AS, Martínez-González J-M, Gómez-Font R et al (2010) Current status of the torus palatinus and torus mandibularis. Medicina Oral, Patologia Oral y Cirugia Bucal 15:e353–e360

Godinho M, Barbosa F, Andrade F et al (2013) Torus palatinus osteonecrosis related to bisphosphonate: a case report. Case Rep Dermatol 5:120–125

Gorsky M, Raviv M, Kfir E et al (1996) Prevalence of torus palatinus in a population of young and adult Israelis. Arch Oral Biol 41:623–625

Haugen LK (1992) Palatine and mandibular tori. A morphologic study in the current Norwegian population. Acta Odontol Scand 50:65–77

Hiremath VK, Husein A, Mishra N (2011) Prevalence of torus palatinus and torus mandibularis among Malay population. J Int Soc Prev Community Dent 1:60–64

Imada TSN, Tjioe KC, Sampieri MB d S et al (2014) Surgical management of palatine Torus – case series. Revista de Odontologia da UNESP. Revista de Odontologia da UNESP/Universidade Estadual Paulista Júlio de Mesquita Filho 43:72–76

Jainkittivong A, Apinhasmit W, Swasdison S (2007) Prevalence and clinical characteristics of oral tori in 1,520 Chulalongkorn University dental school patients. Surg Radiol Anat 29:125–131

Kaneko K, Takahashi H (2014) Bisphosphonate-related osteonecrosis of the palatal torus. ORL 76:353–356

Kang S, Lee S-J, Ahn S-J et al (2007) Bone thickness of the palate for orthodontic mini-implant anchorage in adults. Am J Orthod Dentofac Orthop 131:S74–S81

King DR, Moore GE (1971) The prevalence of torus palatinus. J Oral Med 26:113–115

Marx RE (2003) Pamidronate (Aredia) and zoledronate (Zometa) induced avascular necrosis of the jaws: a growing epidemic. J Oral Maxillofac Surg 61:1115–1117

Miwa Y, Asaumi R, Kawai T et al (2018) Morphological observation and CBCT of the bony canal structure of the groove and the location of blood vessels and nerves in the palatine of elderly human cadavers. Surg Radiol Anat 40:199–206

Moraes Junior EF, Damante CA, Araujo SR (2010) Torus palatinus: a graft option for alveolar ridge reconstruction. Int J Periodontics Restorative Dent 30:283–289

Sathya K, Kanneppady SK, Arishiya T (2012) Prevalence and clinical characteristics of oral tori among outpatients in Northern Malaysia. J Oral Biol Craniofac Res 2:15–19

Sisman Y, Ertas ET, Gokce C et al (2008) Prevalence of torus palatinus in cappadocia region population of Turkey. Eur J Dent 2:269–275

Tomaszewska IM, Tomaszewski KA, Kmiotek EK et al (2014) Anatomical landmarks for the localization of the greater palatine foramen-a study of 1200 head CTs, 150 dry skulls, systematic review of literature and meta analysis. J Anat 225:419–435

Yildiz E, Deniz M, Ceyhan O (2005) Prevalence of torus palatinus in Turkish schoolchildren. Surg Radiol Anat 27:368–371. http://link.springer.com/10.1007/s00276-005-0003-x, [cited 2018 Jan 2]

Yoshinaka M, Ikebe K, Furuya-Yoshinaka M et al (2010) Prevalence of torus palatinus among a group of Japanese elderly. J Oral Rehabil 37:848–853

Part V

Lingual Plate and Oral Floor

Anatomy and Variations of the Submandibular Fossa

13

Yosuke Harazono

13.1 Introduction

The submandibular fossa (Fig. 13.1) is a concavity in the lingual part of the mandibular body, inferior to the mylohyoid line, in the lower molar region, where part of the submandibular gland is situated. Around the submandibular fossa are many anatomical structures implicated in implant surgery, salivary calculus, and static bone cavity. Therefore, dentists need a thorough understanding of the surgical anatomy in the area of the submandibular fossa to preclude complications including hemorrhages, nerve injuries, ductal strictures, and perforation of the lingual cortical bone. In this chapter, the anatomy and its variations in the area of the submandibular fossa are reviewed and clinical considerations are described.

13.2 Sublingual Gland

The sublingual gland is a long flattened body located above the mylohyoid muscle and lateral to the genioglossus muscle (Figs. 13.2 and 13.3). The body of the sublingual gland bulges out the mucosa on the floor of the mouth. The superior border of the sublingual gland represents the sublingual fold in the floor of the mouth, on which the numerous orifices of minor sublingual ducts open (Rivinus's duct) (Amano et al. 2012). The anterior portion of the sublingual gland may contains a single larger secretory duct (Bartholin's duct) that merges with the submandibular duct superior to the hyoglossus muscle. The merged ducts open at the sublingual caruncle.

Y. Harazono (✉)
Department of Dentistry and Oral Surgery, NTT Medical Center, Tokyo, Japan

J. Iwanaga, R. S. Tubbs (eds.), *Anatomical Variations in Clinical Dentistry*,
https://doi.org/10.1007/978-3-319-97961-8_13

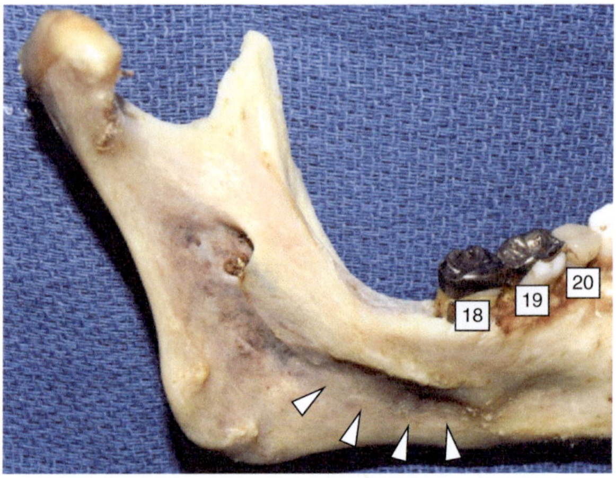

Fig. 13.1 Submandibular fossa (arrowheads)

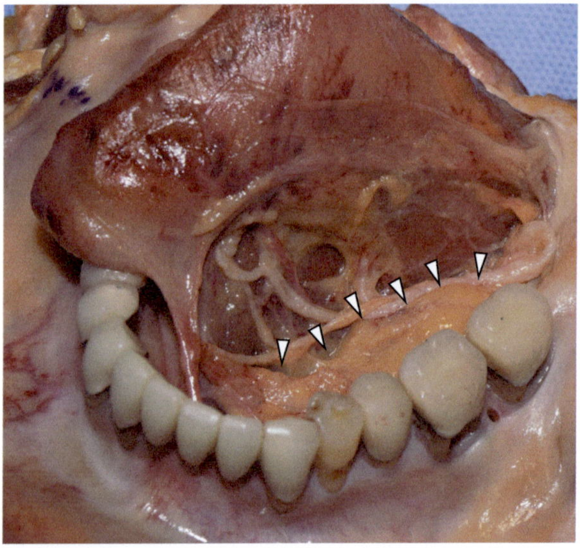

Fig. 13.2 A long and flattened body of the sublingual gland (arrowheads)

13.3 Submandibular Gland

The submandibular gland is a rounded biconvex body occupying the submandibular triangle and is enclosed in a capsule that is, in part, derived from the investing layer of the deep cervical fascia. Its upper pole lies on the medial surface of the mandible in the submandibular fossa. Its lower pole extends beyond the boundaries of the submandibular triangle. The inner surface of the submandibular gland is in contact with the stylohyoid, digastric, and styloglossus muscles posteriorly and the hyoglossus and mylohyoid muscles anteriorly. The submandibular duct (Wharton's duct) extends anteriorly and medially from the upper part of the inner surface of the

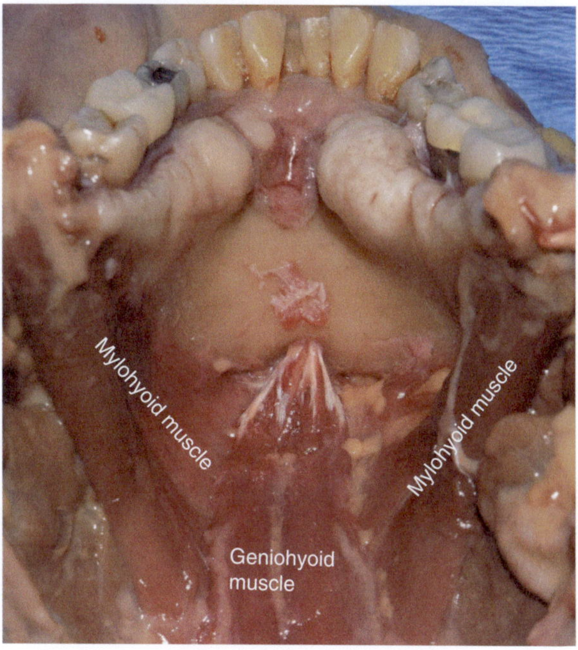

Fig. 13.3 Posterior view of the mylohyoid muscle

submandibular gland. It courses along the inner surface of the sublingual gland after crossing the lingual nerve superiorly and runs forward in the sublingual space to open in the sublingual caruncle with the major sublingual duct (Amano 2011; Williams et al. 1989). The facial artery is close to the surface and the upper border of the submandibular gland, running in a groove of the gland and sometimes even embedded in the glandular body itself. It sends branches into the gland and is thus tightly attached to it.

13.4 Sublingual and Submental Arteries

The lingual artery branches into the sublingual artery at the anterior border of the hyoglossus muscle (He et al. 2017). The sublingual artery travels through the floor of the mouth on the mylohyoid muscle to supply the sublingual gland, the mylohyoid, geniohyoid, and genioglossus muscles, the mucous membranes of the floor of the mouth, and the lingual gingiva (Bavitz et al. 1994; Hofschneider et al. 1999; Kalpidis and Setayesh 2004; Kawai et al. 2006; Yildirim et al. 2014; Wang et al. 2015) (Fig. 13.4a).

The submental artery branches from the facial artery and travels inferior to the mylohyoid muscle along with the nerve to the mylohyoid, supplying the submandibular lymph nodes, the anterior belly of the digastric muscle, and the mylohyoid muscle (He et al. 2017) (Fig. 13.4b).

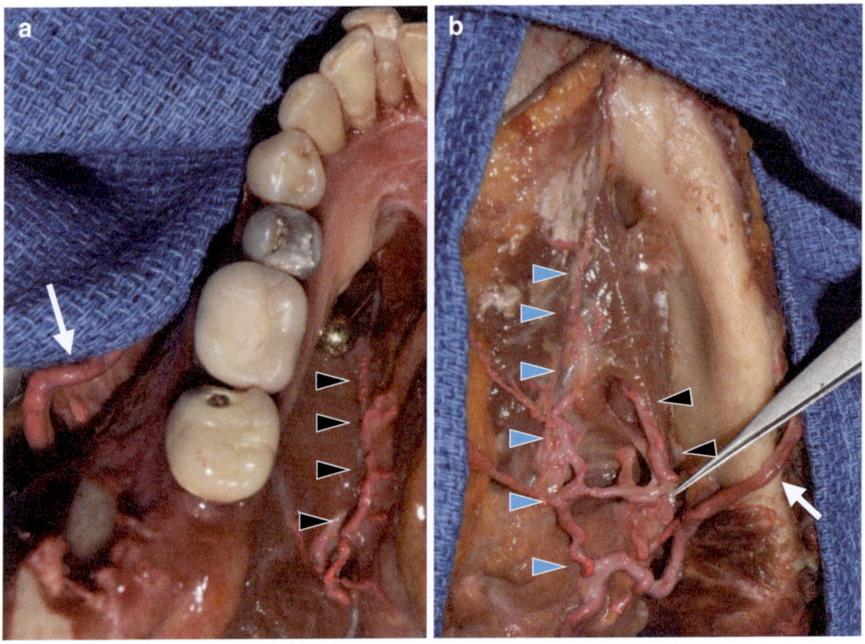

Fig. 13.4 The sublingual artery (black arrowheads) travels above the mylohyoid muscle. The submental artery (blue arrowheads) branches off the facial artery (arrows) and travels below the mylohyoid muscle. (**a**) Superior view. (**b**) Inferior view

Fig. 13.5 Anastomosis of the submental and sublingual arteries

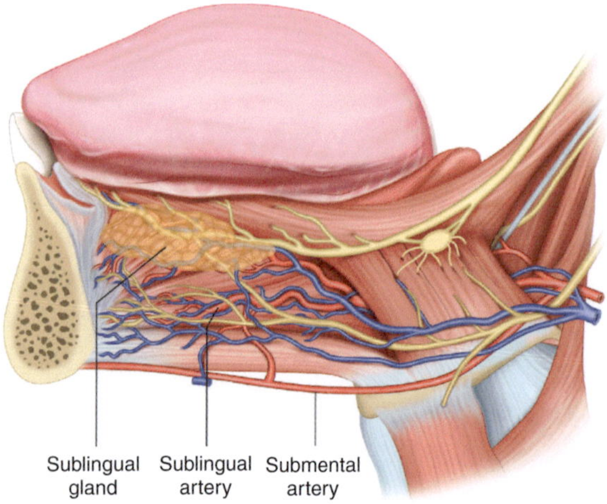

The sublingual and submental arteries anastomose in about 7% of cases via a submental branch penetrating through the posterior mylohyoid muscle to reach the anterior sublingual space (Fig. 13.5). The sublingual artery is generally described as a branch of the lingual artery in anatomical textbooks. However, Nakajima et al. (2014) suggested that the facial artery contributes as much to the composition of the sublingual artery as the lingual artery (Mendes et al. 2014; He et al. 2017).

13.5 Lateral Lingual Foramen

There are two groups of lingual foramina differentiated by their location in the mandible: the median lingual foramen and the lateral lingual foramen. The lateral lingual foramen is most commonly located at the mandibular second premolar, followed by the mandibular first premolar (Sanomiya Ikuta et al. 2016) (Fig. 13.6). Salinas-Goodier et al. (2016) reported that a lingual foramen was the most frequently observed accessory foramen of the mandible, present in 96% of 46 dried mandibles. The branch of the sublingual artery, submental artery and their anastomosis enter the lateral lingual foramen (Nakajima et al. 2014), which is regularly detectable using CBCT; its diameter ranges from 0.25 to 1.6 mm (Wang et al. 2015).

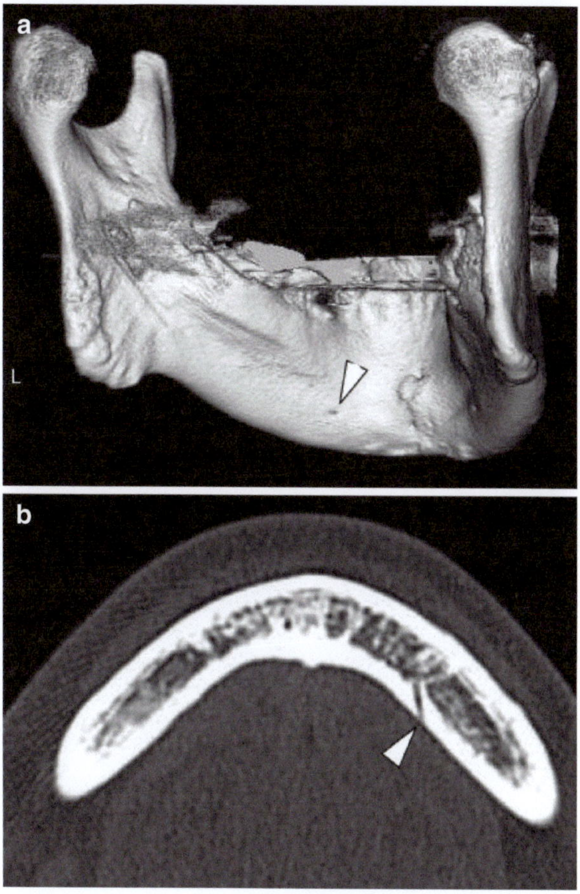

Fig. 13.6 The lateral lingual foramen (arrowheads) detected on computed tomographic images by three-dimensional viewer (**a**) and axial images (**b**)

13.6 Lingual Nerve

The lingual nerve descends between the lateral and medial pterygoid muscles and may be separated from the inferior alveolar nerve by the pterygospinous ligament (Ligament of Civinini) (Peuker et al. 2001)), usually 5–10 mm below the cranial base. Coursing downward and slightly laterally, it follows the lateral surface of the medial pterygoid muscle. At the level of the upper end of the mylohyoid line, it curves sharply anteriorly and continues horizontally on the superior surface of the mylohyoid muscle into the oral cavity. The lingual nerve is closely related to the upper pole of the submandibular gland, where it releases fibers to the submandibular ganglion. Further anteriorly, it supplies the mucous membrane of the sublingual region and enters the substance of the tongue (Fig. 13.7). The anatomical relationship between the lingual nerve and Wharton's duct is quite variable. Mendes et al. reported that Wharton's duct crossed the lingual nerve superiorly in 62.5% of cases and coursed beneath the nerve in 37.5% (Mendes et al. 2014). Similarly, Hölzle and Wolff (2001) reported on four of 34 cases (11.7%) in which Wharton's duct coursed deep to the lingual nerve.

13.7 Clinical Considerations

13.7.1 Implant Surgery

The sublingual fossa is located in the canine and premolar regions (Uchida et al. 2012). This location corresponds to the implant sites in many cases of serious hemorrhage in the floor of the mouth (Mason et al. 1990; Mordenfeld et al. 1997; Kalpidis and Konstantinidis 2005), suggesting that such hemorrhages are very likely to be caused by perforation of the lingual cortical bone followed by injury to the vessels penetrating the lateral lingual foramen (Nakajima et al. 2014) in the canine and premolar regions (Fig. 13.8). Uchida et al. reported that if the lingual

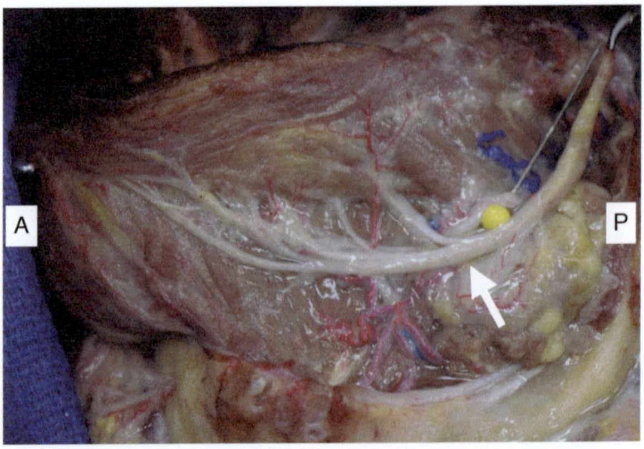

Fig. 13.7 Left sublingual surface. Note that the lingual nerve (arrow) supplies the mucous membrane of the sublingual region and enters the substance of the tongue

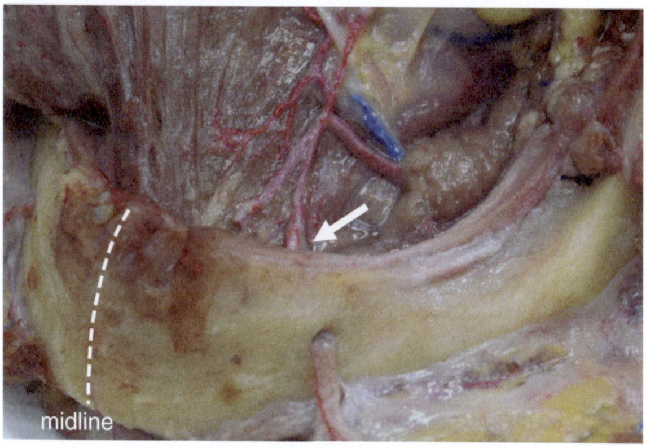

Fig. 13.8 Left oral floor. Note that the vessel penetrates the lateral lingual foramen (arrow) in the canine and premolar regions

cortical bone is perforated near the submandibular fossa, the vessels penetrating the lateral lingual foramen are likely to be injured when an implant with diameter ≥ 3.75 mm and length ≥ 15 mm is placed in the canine and premolar regions (Uchida et al. 2015). To preclude perforation of the lingual cortical bone during implant surgery, comprehensive evaluation of the characteristics of this region is necessary. The lingual flap should be retracted to confirm the shape of the lingual bone and existence of the lingual foramina definitively. Clinicians must beware of vessel injury, especially when there is a lingual foramen larger than 1 mm in diameter (Kim et al. 2013). CT imaging allows the anatomy of the posterior mandible to be characterized and provides information including the degree of lingual concavity in the area of the submandibular fossa (Herranz-Aparicio et al. 2016) or the correlation between submandibular fossa depth and bone thickness (de Souza et al. 2016).

13.7.2 Salivary Calculus

Sialolithiasis varies in size, shape, texture, and consistency; it can be solitary or multiple. Approximately 80% of sialolithiasis involves the submandibular glands, 20% occurs in the parotid gland, and less than 1% is found in the sublingual gland (Witt and Edkins 2017). The conventional treatment has shifted from open surgical or gland resection procedures to endoscopic, endoscopic-assisted, and transoral gland preservation techniques (Witt et al. 2012). Risks associated with transoral submandibular approaches include lingual nerve injury, bleeding, duct perforations, ductal strictures, and development of traumatic ranulas (Walvekar et al. 2008; Bowen et al. 2011; Nahlieli et al. 2003). Although feasible in selected cases, the same sialolithotomy procedure for the more posterior ductal region entails a significant risk of injury to the lingual nerve and the possibility of severe bleeding from the lingual vessels (Witt and Edkins 2017). We need to understand the surgical anatomy in the area of submandibular fossa to avoid these complications (Fig. 13.9).

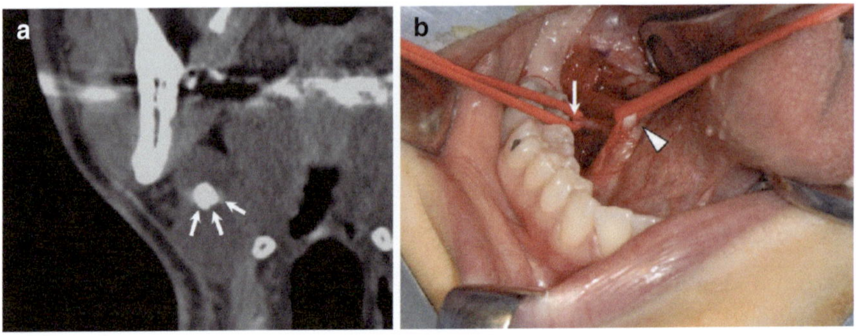

Fig. 13.9 (**a**) Salivary calculus on computed tomographic image within hilum of submandibular gland. (**b**) Operative finding in intraoral submandibular sialolithotomy procedure (arrow: Wharton's duct, arrowhead: lingual nerve)

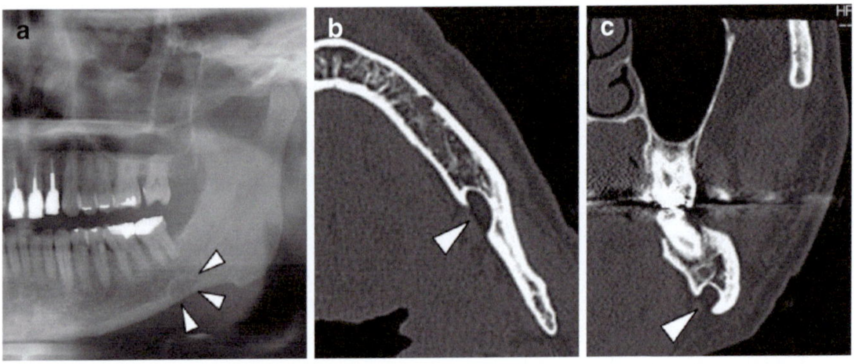

Fig. 13.10 Static bone cavity (arrowheads) detected in panoramic radiography image (**a**) and computed tomographic images (**b**: axial section, **c**: coronal section)

13.7.3 Static Bone Cavity

Static bone cavity is a localized bone defect predominantly affecting the mandible. The following description of these bone cavities remains in use today: a round or oval structure with a diameter of 5–30 mm, mostly located beneath the mandibular canal (Stafne 1942). The etiology of static bone cavity is unknown. One speculation involves atrophy of the mandible induced by pressure of the dorsal part of the submandibular gland and displacement of the sublingual gland (Tolman and Stafne 1967). Radiological examination using panoramic views plus further diagnostics such as CT or MRI are sufficient, and the incidence ranges from 0.08 to 0.7% (Fig. 13.10). According to the retrospective study by Assaf et al. (2014), static bone cavity was found in the mandibular corpus in 72.7% of cases and in the mandibular angle in 27.3%. If static bone cavity is developing, a biopsy can aid the differential diagnosis from cystic or tumor lesions, and long term follow-ups are strongly recommended (Karube et al. 2016).

References

Amano O (2011) The salivary gland: anatomy for surgeons and researchers. Jpn J Oral Maxillofac Surg 57:384–393

Amano O, Mizobe K, Bando Y et al (2012) Anatomy and histology of rodent and human major salivary glands: overview of the Japan salivary gland society-sponsored workshop. Acta Histochem Cytochem 45:241–250. https://doi.org/10.1267/ahc.12013

Assaf AT, Solaty M, Zrnc TA et al (2014) Prevalence of Stafne's bone cavity–retrospective analysis of 14,005 panoramic views. In Vivo 28:1159–1164

Bavitz JB, Harn SD, Homze EJ (1994) Arterial supply to the floor of the mouth and lingual gingiva. Oral Surg Oral Med Oral Pathol 77:232–235

Bowen MA, Tauzin M, Kluka EA et al (2011) Diagnostic and interventional sialendoscopy: a preliminary experience. Laryngoscope 121:299–303. https://doi.org/10.1002/lary.21390

de Souza LA, Souza Picorelli Assis NM, Ribeiro RA et al (2016) Assessment of mandibular posterior regional landmarks using cone-beam computed tomography in dental implant surgery. Ann Anat 205:53–59. https://doi.org/10.1016/j.aanat.2016.01.006

Harry S, DuBrul EL (1988) Sicher and DuBrul's oral anatomy, 8th edn. IshiyakuEuroAmerica, St. Louis; Tokyo

He P, Truong MK, Adeeb N et al (2017) Clinical anatomy and surgical significance of the lingual foramina and their canals. Clin Anat 30:194–204. https://doi.org/10.1002/ca.22824

Herranz-Aparicio J, Marques J, Almendros-Marqués N et al (2016) Retrospective study of the bone morphology in the posterior mandibular region. Evaluation of the prevalence and the degree of lingual concavity and their possible complications. Med Oral Patol Oral Cir Bucal 21:e731–e736

Hofschneider U, Tepper G, Gahleitner A et al (1999) Assessment of the blood supply to the mental region for reduction of bleeding complications during implant surgery in the interforaminal region. Int J Oral Maxillofac Implants 14:379–383

Hölzle FW, Wolff KD (2001) Anatomic position of the lingual nerve in the mandibular third molar region with special consideration of an atrophied mandibular crest: an anatomical study. Int J Oral Maxillofac Surg 30:333–338

Kalpidis CD, Konstantinidis AB (2005) Critical hemorrhage in the floor of the mouth during implant placement in the first mandibular premolar position: a case report. Implant Dent 14:117–124

Kalpidis CD, Setayesh RM (2004) Hemorrhaging associated with endosseous implant placement in the anterior mandible: a review of the literature. J Periodontol 75:631–645

Karube T, Yasui T, Onizawa K et al (2016) A long-term followed-up case of a developing lingual bone defect associated with the sublingual gland. Jpn J Oral Maxillofac Surg 62:299–303

Kawai T, Sato I, Yosue T et al (2006) Anastomosis between the inferior alveolar artery branches and submental artery in human mandible. Surg Radiol Anat 28:308–310

Kim DH, Kim MY, Kim CH (2013) Distribution of the lingual foramina in mandibular cortical bone in Koreans. J Korean Assoc Oral Maxillofac Surg 39:263–268. https://doi.org/10.5125/jkaoms.2013.39.6.263

Mason ME, Triplett RG, Alfonso WF (1990) Life-threatening hemorrhage from placement of a dental implant. J Oral Maxillofac Surg 48:201–204

Mendes MB, de CarvalhoLeite Leal Nunes CM, de Almeida Lopes MC (2014) Anatomical relationship of lingual nerve to the region of mandibular third molar. J Oral Maxillofac Res 4:e2. https://doi.org/10.5037/jomr.2013.4402

Mordenfeld A, Andersson L, Bergström B (1997) Hemorrhage in the floor of the mouth during implant placement in the edentulous mandible: a case report. Int J Oral Maxillofac Implants 12:558–561

Nahlieli O, Shacham R, Bar T et al (2003) Endoscopic mechanical retrieval of sialoliths. Oral Surg Oral Med Oral Pathol Oral Radiol Endod 95:396–402

Nakajima K, Tagaya A, Otonari-Yamamoto M et al (2014) Composition of the blood supply in the sublingual and submandibular spaces and its relationship to the lateral lingual foramen of the

mandible. Oral Surg Oral Med Oral Pathol Oral Radiol 117:e32–e38. https://doi.org/10.1016/j.oooo.2012.03.032
Paulsen F, Waschke J (2011) Sobotta atlas of human anatomy -head, neck, and neuroanatomy, 15th edn. Urban & Fischer, Munich
Peuker ET, Fischer G, Filler TJ (2001) Entrapment of the lingual nerve due to an ossified pterygospinous ligament. Clin Anat 14:282–284
Salinas-Goodier C, Manchón A, Rojo R et al (2016) Prevalence and location of accessory foramina in the human mandible. Oral Radiol 32:72–78
Sanomiya Ikuta CR, Paes da Silva Ramos Fernandes LM, Poleti ML et al (2016) Anatomical study of the posterior mandible: lateral lingual foramina in cone beam computed tomography. Implant Dent 25:247–251. https://doi.org/10.1097/ID.0000000000000387
Stafne E (1942) Bone cavities situated near the angle of the mandible. J Am Dent Assoc 29:1969–1972
Tolman DE, Stafne EC (1967) Developmental bone defects of the mandible. Oral Surg Oral Med Oral Pathol 24:488–490
Uchida Y, Goto M, Danjo A et al (2012) Anatomic measurement of the depth and location of the sublingual fossa. Int J Oral Maxillofac Surg 41:1571–1576. https://doi.org/10.1016/j.ijom.2012.03.015
Uchida Y, Goto M, Danjo A et al (2015) Anatomical relationship between the sublingual fossa and the lateral lingual foramen. Int J Oral Maxillofac Surg 44:1146–1151. https://doi.org/10.1016/j.ijom.2015.04.004
Walvekar RR, Razfar A, Carrau RL et al (2008) Sialendoscopy and associated complications: a preliminary experience. Laryngoscope 118:776–779. https://doi.org/10.1097/MLG.0b013e318165e355
Wang YM, Ju YR, Pan WL et al (2015) Evaluation of location and dimensions of mandibular lingual canals: a cone beam computed tomography study. Int J Oral Maxillofac Surg 44:1197–1203. https://doi.org/10.1016/j.ijom.2015.03.014
Williams PL, Warwick R, Dyson M et al (1989) Gray's anatomy, 37th edn. Churchill Livingstone, Edinburgh, pp 1290–1298
Witt RL, Edkins O (2017) Sialolithiasis: traditional & sialendoscopic techniques. The open access atlas of otolaryngology, head & neck operative surgery
Witt RL, Iro H, Koch M et al (2012) Minimally invasive options for salivary calculi. Laryngoscope 122:1306–1311. https://doi.org/10.1002/lary.23272
Yildirim YD, Güncü GN, Galindo-Moreno P et al (2014) Evaluation of mandibular lingual foramina related to dental implant treatment with computerized tomography: a multicenter clinical study. Implant Dent 23:57–63. https://doi.org/10.1097/ID.0000000000000012

Anatomy and Variations of the Sublingual Space

14

Norie Yoshioka

14.1 Introduction

The sublingual space is superomedial to the mylohyoid muscle and lateral to the genioglossus and geniohyoid muscles in the oral cavity. Its major contents are the sublingual salivary gland and its ducts (the major sublingual duct, Bartholin's duct; the smaller sublingual ducts, Rivinus's ducts), the submandibular duct (Wharton's duct), the superior portion of the submandibular salivary gland, the lingual nerve, artery and vein, and the glossopharyngeal (IX) and hypoglossal (XII) nerves. The sublingual space communicates with the submandibular space at the posterior margin of mylohyoid muscle where there is a gap between this muscle and hyoglossus muscle (Otonari-Yamamoto et al. 2010). Many lesions such as ranula and submandibular duct obstruction, various malignancies, inflammation, and vascular abnormalities arise uniquely in the sublingual space. Better knowledge of the complex muscular, vascular, glandular, ductal, and neural anatomy of this region is important for accurate diagnosis and treatment planning by both general dentists and oral surgeons in order to avoid unnecessary complications. In this chapter, the normal anatomy and variations of the sublingual space are reviewed and its clinical relevance is discussed.

N. Yoshioka (✉)
Department of Oral and Maxillofacial Surgery, Okayama University Graduate School of Medicine, Dentistry and Pharmaceutical Sciences, Okayama, Japan
e-mail: noriy@md.okayama-u.ac.jp

J. Iwanaga, R. S. Tubbs (eds.), *Anatomical Variations in Clinical Dentistry*,
https://doi.org/10.1007/978-3-319-97961-8_14

14.2 Sublingual Gland and Its Ducts

The floor of the mouth contains several salivary glands including the sublingual glands, the superior portion of the submandibular glands, and the subepithelial minor salivary glands. The sublingual glands, which are almond-shaped, are the smallest of the three paired major salivary glands. They are both situated under the tongue and bordered laterally by the mandible and medially by the genioglossus muscle. Secretions drain into the oral cavity via the minor sublingual ducts (Rivinus's ducts), of which there are 8–20 per gland, each opening on to the sublingual folds (Fig. 14.1). Occasionally, several of the more anterior ducts join to form a common duct (Bartholin's duct), which typically empties into the main submandibular duct. Bartholin's duct arises from the inferior aspect of the sublingual gland and then adheres to the passing submandibular duct on its medial side.

14.3 Submandibular Duct

The submandibular gland has a secretory duct named Wharton's duct, which is 4–5 cm long and 1.5 mm in diameter. Wharton's duct has a pathway that emerges from the medial part of the medial face of the gland, crossing inward the space between the hyoglossus muscle medially and the mylohyoid muscle laterally, to end in the sublingual caruncle, which is located in the floor of the mouth, to the sides of the lingual frenulum, through perforation of the mucous layer on both sides (Buckenham 2004; Manzur-Villalobos et al. 2016). There are a few reports of anatomical variations of Wharton's duct. Manzur-Villalobos et al. (2016) reported that Wharton's duct joined and anastomosed with the secretory duct of the parotid gland, called Stensen's duct, in a black male (Otonari-Yamamoto et al. 2010). Gadodia et al. (2007) described an extremely rare case of an accessory submandibular gland and duct: there was a unilateral duplication of the submandibular duct with a small duct, approximately 2 cm long, running parallel to the main duct and draining a small accessory gland (Saigusa et al. 2006).

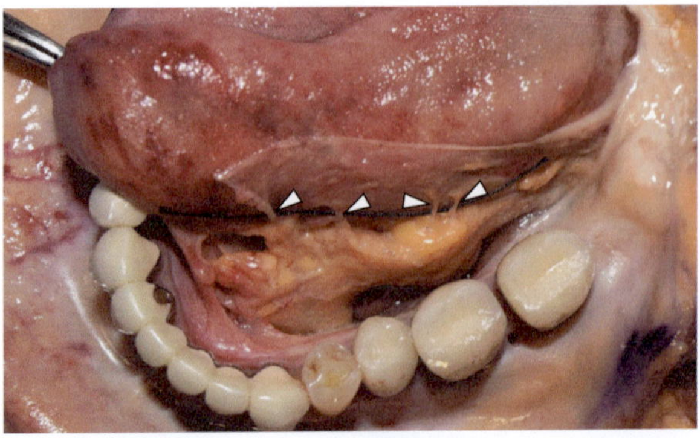

Fig. 14.1 Opening of minor sublingual ducts (Rivinus's ducts)

14.4 Lingual Nerve and Its Relationship to the Submandibular Duct

The lingual nerve (LN) is one of the two terminal branches of the posterior division of the mandibular nerve. It supplies general sensation to the mucosa of the anterior two-thirds of the tongue, the sublingual mucosa, the mandibular lingual gingiva, and the floor of the mouth. It usually passes lateral to the submandibular duct, winds below it, and then passes upward and forward on its medial side on its way to the surface of hyoglossus (Fig. 14.2) (Erdogmus et al. 2008; Otonari-Yamamoto et al. 2010). However, Mendes et al. (2013) reported that this relationship only occurred in 62.5% of cadaver head sides, the remaining 37.5% cases showing the LN crossing above the duct (Gadodia et al. 2007). Occasionally, the submandibular duct runs deep in the floor of the mouth with no relationship to the LN. In Hölzle's cadaveric study (2001), this occurred in 11.8% of the dissections (Gadodia et al. 2007). No constant position has been found to relate the point of intersection, as it can happen anywhere from the distal part of the second premolar tooth to the retromolar triangle. As for the overlap between the LN and the submandibular duct, Al-Amery et al. (2016) reported that the loop of the LN started around the third molar region in 76.9%, but all looping ended at around the first and second molar region in seven Asian cadavers (Buckenham 2004). Awareness of the variations of the LN is important for preventing complications or nerve injury during surgical procedures related to the alveolar crest, submandibular gland/duct, and surrounding areas.

14.5 Anastomosis Between Lingual and Hypoglossal Nerves: Hypoglossal-Lingual Communication

There are reports regarding the communication between the LN and the hypoglossal nerve (HN). Saigusa et al. (2006) reported extra-lingual connections between these two nerves in 83%, and the branch of the HN connected to the LN in the inferior longitudinal muscles of 12 human adult tongue specimens

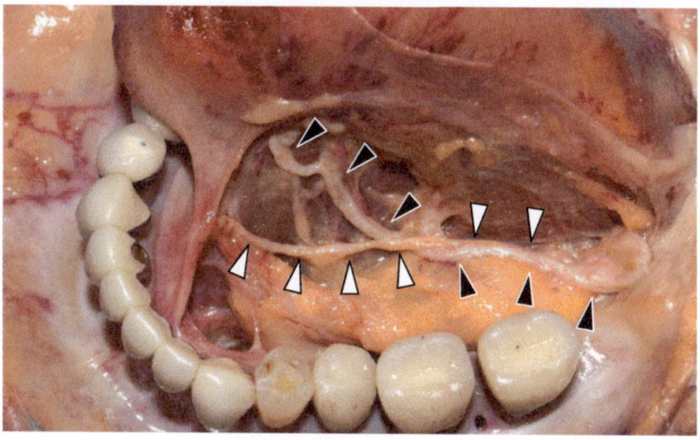

Fig. 14.2 Submandibular duct (white arrowheads) and lingual nerve (black arrowheads)

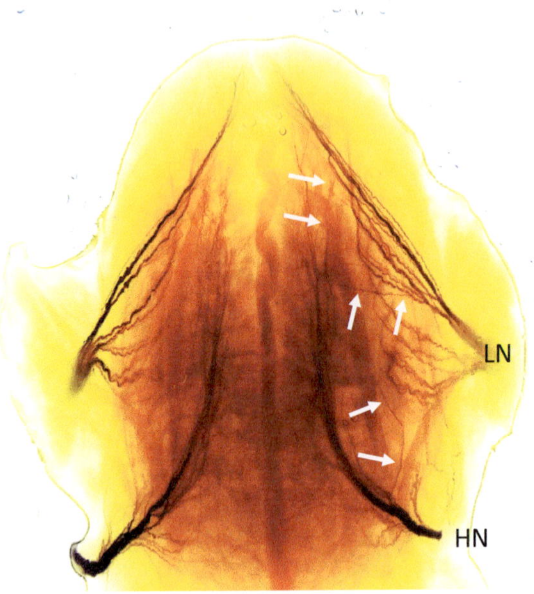

Fig. 14.3 Sihler's staining in human cadaveric tongue. Note the lingual nerve (LN) and hypoglossal nerve (HN) have communicating branches (arrows)

(Fig. 14.1). Păduraru et al. (2013) also reported intralingual neural anastomoses between the LN and HN. They found several patterns of such anastomoses in ten human adult tongues and a pediatric one: (1) longitudinal anastomoses of the lateral trunks of the HN and LN on the outer surface of the hyoglossus muscle; (2) complex, plexiform anastomoses joining the LN and the medial trunk of the HN to the periarterial plexus of the lingual artery; (3) the medial trunk of the HN and the LN branches linked by successive anastomoses, which crossed the deep artery of the tongue ventrally; and (4) anastomoses between successive lingual branches of the LN. Iwanaga et al. (2017), using modified Sihler's staining on human cadaveric tongues, demonstrated communicating branches between the LN and the HN in three parts of the tongue ipsilaterally, namely, at the anterior, the middle, and the posterior of the anterior two-thirds of the tongue (Fig. 14.3) (Hölzle et al. 2001).

14.6 Sialolithiasis of Wharton's Duct and Surgical Technique of Sialolithotomy

Sialolithiasis is the most common disease of the salivary glands, characterized by the development of salivary stones known as calculi or sialoliths in the salivary ducts or in the affected gland. Most salivary stones occur in the submandibular gland (SMG) and its ductal system. Their size can range from less than 1 mm to a few cm in largest diameter, most of them being <10 mm. Giant sialoliths are defined as those exceeding 15 mm in size (Ledesma-Montes et al. 2007; Otonari-Yamamoto et al. 2010). Approximately 40% of SMG stones are located in the distal region of the duct and can be removed easily through an intraoral procedure. However, intraglandular stones and those in the proximal region of the duct are more difficult to remove transorally because of their positions deep in the floor of the mouth. Thus, intraglandular stones or SMG atrophy caused by stones are generally removed via a transcervical approach (McGurk et al. 2005).

14.6.1 Surgical Procedure of Intraoral Stone Removal

An incision is made through the mucosa of the lateral part of the floor of the mouth, extending from the mucosa of the orifice of Wharton's duct to the lingual side of the teeth. Careful dissection is performed between Wharton's duct and the sublingual gland. After the stone is located by bimanual palpation, the duct is incised and the stone is removed. The duct is irrigated with normal saline to clean the region and remove stone debris. A drain gauze is then inserted into the submandibular duct and the incised mucosa of the floor of the mouth is sutured back (Fig. 14.4). When the stone is located in the proximal part of Wharton's duct, transcervical pressure in the submandibular triangle is especially helpful for delivering it on to the floor of the mouth. Also, it is important to identify and dissect the LN gently in order to avoid LN paralysis during the procedure (Fig. 14.5). Intraoral removal can be performed regardless of location if the stone is palpable.

14.6.2 Surgical Procedure of Transcervical Approach to Remove SMG

A 4–6 cm incision is placed in the lateral neck crease approximately 2–3 cm below the lower edge of the mandible (Fig. 14.6). The subplatysmal flap is elevated and the marginal mandibular branch of the facial nerve is identified using a nerve stimulator and protected. When the marginal mandibular branch of the facial nerve is difficult to find, an alternative way to protect it is to identify and ligate the facial vein and then reflect it superiorly with the fascia over the SMG. The facial artery is ligated or can be preserved by ligating only the branches of the facial artery to the SMG. Blunt dissection continues toward the superomedial part of the gland, at which point the mylohyoid muscle must be retracted anteriorly to complete the dissection. Posterior and inferior traction on the SMG facilitates the identification and differentiation of Wharton's duct, the LN with its attachment to the submandibular ganglion, and the HN. After Wharton's duct has been ligated, the SMG is liberated from the submandibular ganglion and removed with preservation of the LN and HN.

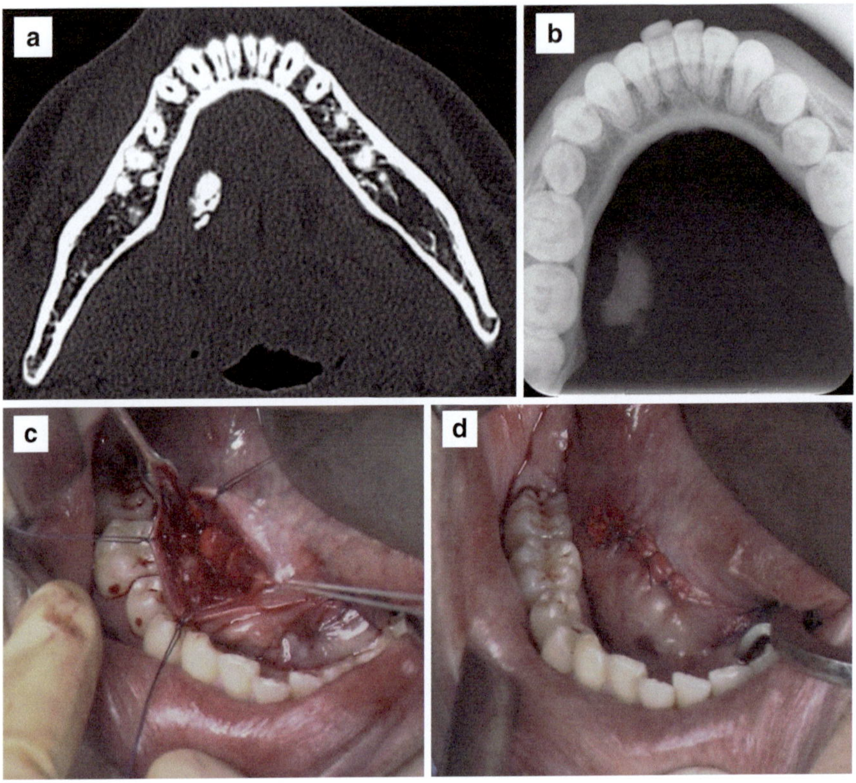

Fig. 14.4 Intraoral surgical procedure for stone removal. (**a**) CT image. (**b**) Occlusal radiograph. (**c**) Detection of the Wharton's duct and the stone. (**d**) The incised mucosa is sutured back after insertion of drain gauze

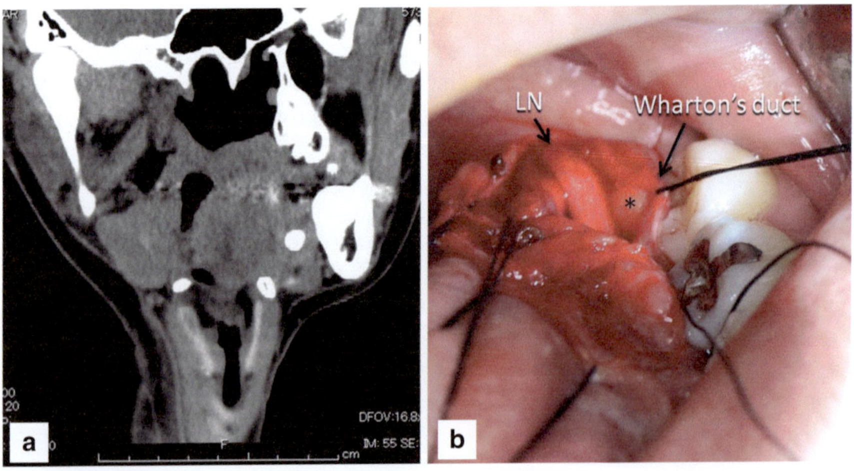

Fig. 14.5 Relationship of the lingual nerve (LN), Wharton's duct and stone (star). (**a**) CT image. (**b**) The lingual nerve runs medial to the Wharton's duct, which descends toward the deep portion of the gland

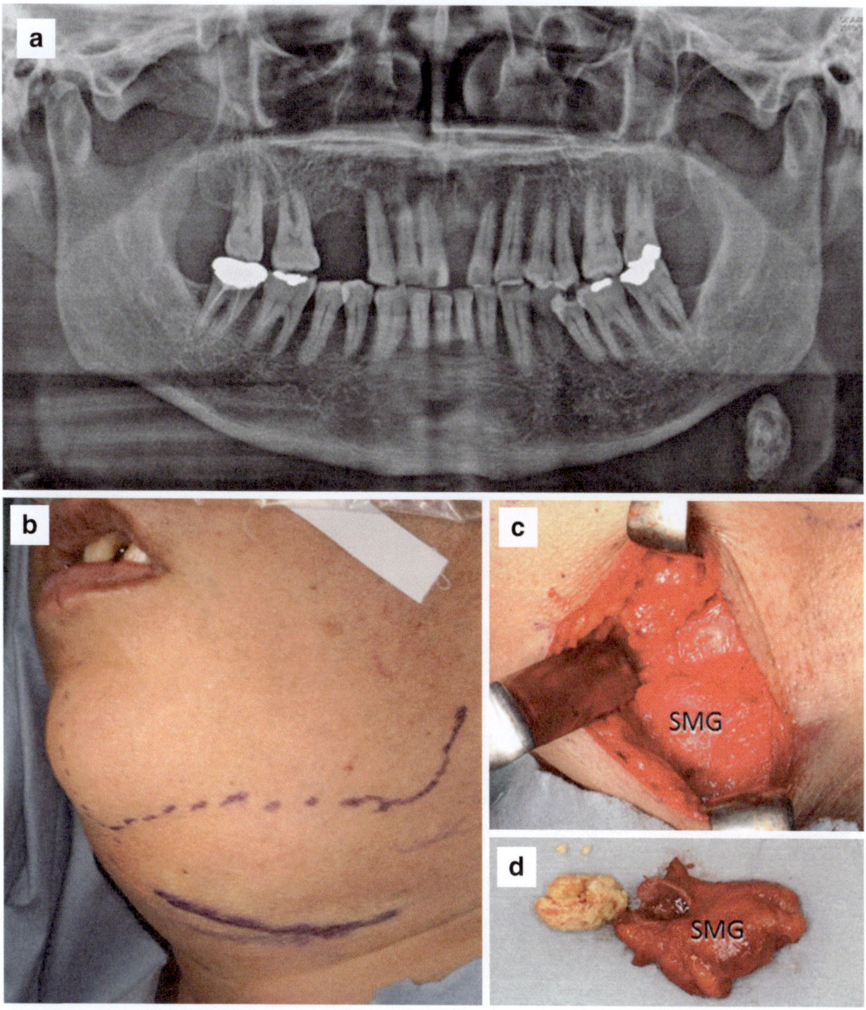

Fig. 14.6 Extraoral surgical procedure of stone removal (transcervical approach). (**a**) Panoramic radiograph. A large radio-opaque mass in the left submandibular gland (SMG). (**b**) Skin incision line. (**c**) Exposure of SMG. (**d**) Stone and SMG

14.7 Abscess Drainage in the Sublingual Space

The sublingual spaces are bounded anteriorly and laterally by the mandible, superiorly by the floor of the mouth and tongue, and inferiorly by the mylohyoid muscle. There is no anatomical barrier between the left and right sublingual spaces. Infection of the sublingual space appears clinically as erythematous, tender swelling of the floor of the mouth, and elevation of the tongue, which causes drooping and dysphagia. The sublingual space can be surgically drained intraorally. The

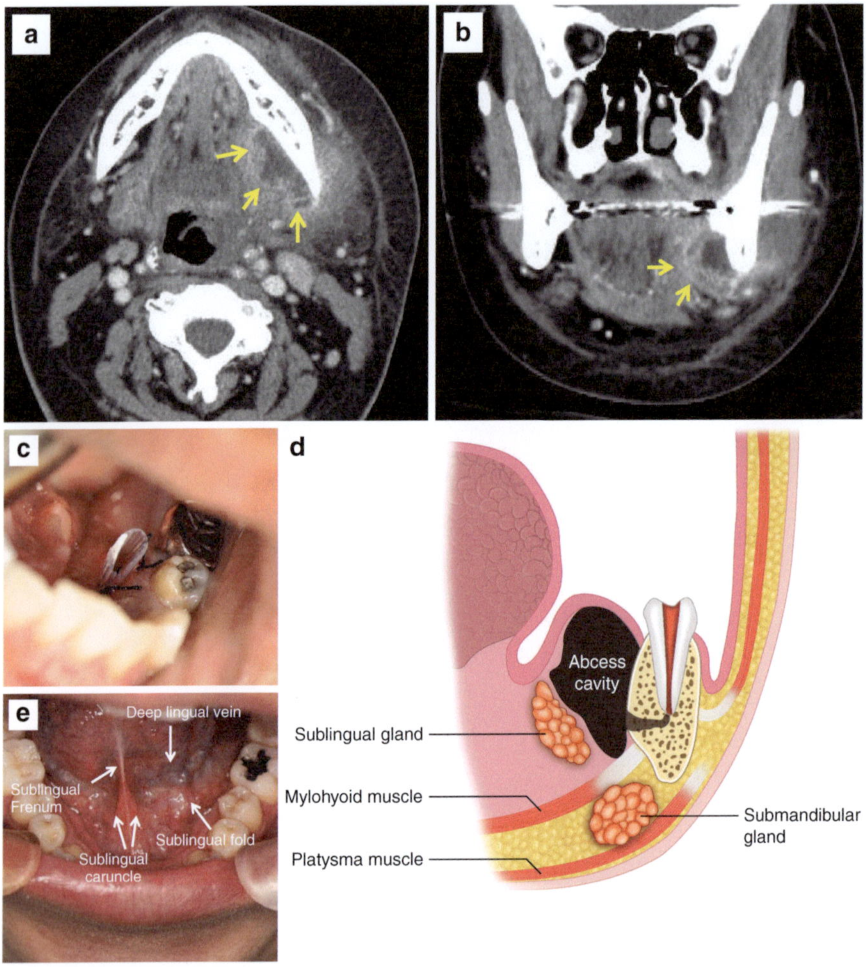

Fig. 14.7 Sublingual abscess drainage. (**a**, **b**) CT image. Arrows; sublingual abscess. (**c**) Insertion of a silicon tube for drainage. (**d**) Schema of sublingual abscess. (**e**) The superficial anatomy of the floor of the mouth

incision should be between lateral to the sublingual fold and medial to the mandibular body and parallel to the mandibular arch. After the mucosa is cut, dissecting bluntly into the abscess, a silicon tube is placed in the abscess cavity for drainage (Fig. 14.7).

References

Al-Amery SM, Nambiar P, Naidu M et al (2016) Variation in lingual nerve course: a human cadaveric study. PLoS One 11:e0162773

Buckenham T (2004) Salivary duct intervention. Semin Intervent Radiol 21:143–148

Erdogmus S, Govsa F, Celik S (2008) Anatomic position of the lingual nerve in the mandibular third molar region as potential risk factors for nerve palsy. J Craniofac Surg 19:264–270
Gadodia A, Seith A, Neyaz Z et al (2007) Magnetic resonance identification of an accessory submandibular duct and gland: an unusual variant. J Laryngol Otol 121:e18.1–e18.3
Hölzle FW, Wolff KD (2001) Anatomic position of the lingual nerve in the mandibular third molar region with special consideration of an atrophied mandibular crest: an anatomical study. Int J Oral and Maxillofac Surg 30:333–338
Iwanaga J, Watanabe K, Saga T et al (2017) Communicating branches between lingual and hypoglossal nerve: observation using Sihler's staining technique. Surg Radiol Anat 39:741–745
Ledesma-Montes C, Garcés-Ortíz M, Salcido-García JF et al (2007) Giant sialolith: case report and review of the literature. J Oral Maxillofac Surg 65:128–130
Manzur-Villalobos I, Pérez-Bula L, Fang L (2016) Anatomical variation of submandibular gland duct. Sch J Dent Sci 3:12–14
McGurk M, Escudier MP, Brown JE et al (2005) Modern management of salivary calculi. Br J Surg 92:107–112
Mendes BMM, de Carvalho Leite Leal Nunes CM, de Almeida Lopes MC (2014) Anatomical relationship of lingual nerve to the region of mandibular third molar. J Oral Maxillofac Res 4:e2.1–e2.7
Otonari-Yamamoto M, Nakajima K, Tsuji Y et al (2010) Imaging of the mylohyoid muscle: separation of submandibular and sublingual spaces. AJR Am J Roentgenol 194:431–438
Păduraru D, Rusu MC (2013) The anatomy of the intralingual neural interconnections. Surg Radiol Anat 35:457–462
Saigusa H, Tanuma K, Yamashita K et al (2006) Nerve fiber analysis for the lingual nerve of the human adult subjects. Surg Radiol Anat 28:59–65

Anatomy and Variations of the Lingual Frenum and Sublingual Surface

15

Shogo Kikuta and Soichiro Ibaragi

15.1 Introduction

The tongue is dynamically related to various oral functions such as speech, breathing, swallowing, and breastfeeding. It is guided forward by the lingual frenum (LF) during growth (Schoenwolf et al. 2015). The frenum is a single fold that tenses between the sublingual surface and the lingual alveolar mucosa and consists of connective tissue covered by the mucosa without major nerves or vessels. Structural disorders and abnormal attachment of the LF can result in ankyloglossia, a congenital oral anomaly commonly known as "tongue-tie." Severe ankyloglossia can entail various functional disorders of the tongue. The sublingual glands located in the sublingual space sometimes induce mucoceles. In this chapter, the normal anatomy and variations of the LF and sublingual surface are reviewed and their clinical relevance is discussed.

15.2 Lingual Frenum

The LF is normally a single fold that tenses between the sublingual surface and the lingual alveolar mucosa. It consists of connective tissue covered with mucosa without major vessels (Fig. 15.1). Although the frenum affects the diagnosis and treatment of various diseases, e.g., ankyloglossia, there are few anatomical studies of its

S. Kikuta (✉)
Kurume University School of Medicine, Dental and Oral Medical Center, Kurume, Fukuoka, Japan

S. Ibaragi
Okayama University Graduate School of Medicine, Dentistry and Pharmaceutical Sciences, Okayama, Japan
e-mail: sibaragi@md.okayama-u.ac.jp

J. Iwanaga, R. S. Tubbs (eds.), *Anatomical Variations in Clinical Dentistry*,
https://doi.org/10.1007/978-3-319-97961-8_15

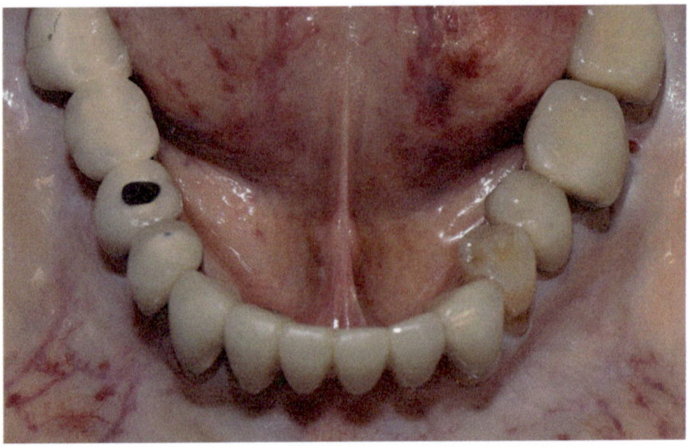

Fig. 15.1 Lingual frenum

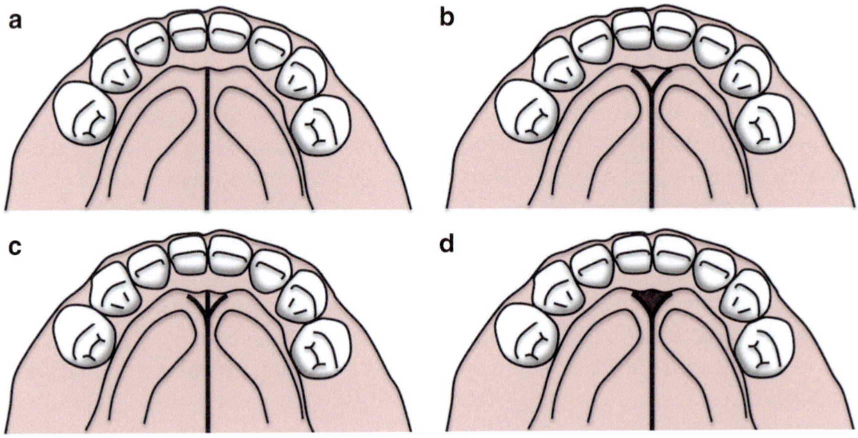

Fig. 15.2 Illustration showing attachment forms of the lingual frenum. (**a**) Type I, (**b**) Type II, (**c**) Type III, (**d**) Type IV

attachment. Murakami (1978) investigated the attachment in 302 adults (146 males and 156 females): 30.13% (91/302) of the attachments were on the sublingual caruncle and 69.87% (211/302) on the alveolar mucosa. Murakami classified the attachments of the frenum into four types depending of the number of strips: one strip (type I), two strips (type II), three strips (type III), and membranous strips (type IV) (Fig. 15.2). Type II was observed in 20.86%, followed by type I in 17.55%, type III in 17.22%, and type IV in 14.28%. The prevalence of type I increases with age while type II decreases. The most frequent position of the lateral edge of the frenum was either on the left or right lateral incisors in 36.09%, followed by the midline of the alveolar mucosa in 17.55%, the canine in 9.27%, and the central incisor in 6.95%.

The LF sustains a balance between the tongue, lip, and growing facial bones during fetal development. After birth, as the tongue muscles lengthen, the LF

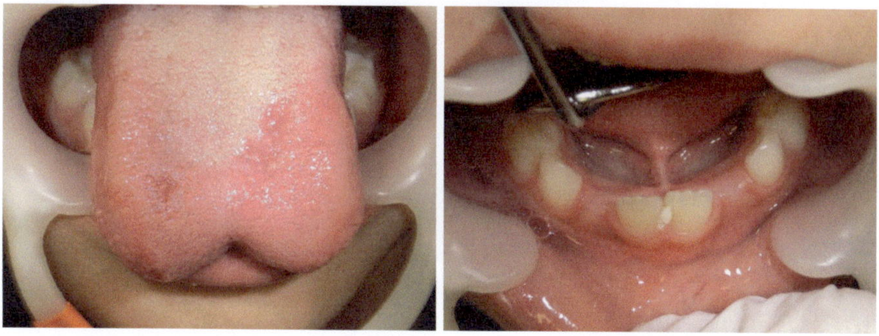

Fig. 15.3 Ankyloglossia known as "tongue-tie" in a child. Heart-shaped tongue during protrusion induced by a short lingual frenum with anterior insertion

contracts and becomes thinner (Tsaousoglou et al. 2015). Anomalies of the LF mostly result from structural disorders. Its failure to recede results in congenital ankyloglossia, commonly known as "tongue-tie" (Fig. 15.3), characterized by an abnormal short and thick LF that restricts tongue mobility (Messner et al. 2000). Several investigators have proposed different classifications for ankyloglossia. Horton et al. (1969) investigated 102 ankyloglossia cases and classified them into the following 3 types: mild type, mucous membrane band; moderate type, markedly fibrosed frenum and genioglossus muscle; and complete type, tongue fused to the floor of the mouth. Kotlow (1999) measured the length of the movable tongue (from the insertion of the LF into the base of the tongue to the tip of tongue) in 322 cases: normal, >16 mm; mild (class I), 12–16 mm; moderate (class II), 8–11 mm; severe (class III), 3–7 mm; and complete (class 4), <3 mm. Several different classifications have been proposed, but none is universally accepted (García Pola et al. 2002; Ruffoli et al. 2005). The prevalence of ankyloglossia ranges from 4.4 to 4.8% in newborns, and the male to female ratio is 3:1 (Messner et al. 2000; Friend et al. 1990; Heller et al. 2005). Although ankyloglossia mostly appears as a sporadic anomaly, it sometimes accompanies other congenital anomalies such as the cleft lip and palate (Klockars and Pitkaranta 2009; Suter and Bornstein 2009). Genetic ankyloglossia develops more commonly in male children, and a relationship to the X chromosome has been suggested (Klockars and Pitkaranta 2009; Marcano et al. 2004; Pauws et al. 2009). Although several diagnostic criteria for the movable range and form of the tongue have been suggested, no consensus has been achieved. The Hazelbaker Assessment Tool for Lingual Frenum Function (HATLFF) defined the severity of ankyloglossia in newborns (Hazelbaker 1993) (Table 15.1). Although ankyloglossia is seen in neonates, children, and adults, it is often asymptomatic. It can result in various problems, e.g., difficulties in breastfeeding, articulatory and speech disorders, poor oral hygiene, and bullying during childhood and adolescence (Segal et al. 2007).

Table 15.1 Hazelbaker Assessment Tool for Lingual Frenum Function (HATLFF) (Hazelbaker 1993)

Lingual function scores	Lingual appearance scores
Lateralization 2: Complete 1: Body of tongue but not tongue tip 0: None Lift of tongue 2: Tip to mid-mouth 1: Only edges to mid-mouth 0: Tip stays at lower alveolar ridge or rises to mid-mouth only with jaw closure Extension of tongue 2: Tip over lower lip 1: Tip over lower gum only 0: Neither of the above, or anterior or mid-tongue humps Spread of anterior tongue 2: Complete 1: Moderate or partial 0: Little or none Cupping 2: Entire edge, firm cup 1: Side edges only, moderate cup 0: Poor or no cup Peristalsis 2: Complete, anterior to posterior 1: Partial, originating posterior to tip 0: None or reverse motion Snapback 2: None 1: Periodic 0: Frequent or with each suck	Appearance of tongue when lifted 2: Round or square 1: Slight cleft in tip apparent 0: Heart- or V-shaped Elasticity of frenum 2: Very elastic 1: Moderately elastic 0: Little or no elasticity Length of lingual frenulum when tongue lifted 2: >1 cm 1: 1 cm 0: <1 cm Attachment of lingual frenulum to tongue 2: Posterior to tip 1: At tip 0: Notched tip Attachment of lingual frenulum to inferior alveolar ridge 2: Attached to floor of mouth or well below ridge 1: Attached just below ridge 0: Attached at ridge

15.2.1 Ankyloglossia and Breastfeeding in Neonates

The most common problems of a baby with ankyloglossia before frenectomy are difficulty in latching on to the breast and sore nipples (Messner et al. 2000; Hogan et al. 2005; Amir et al. 2005; Wallace and Clarke 2006). In one randomized control study, the breastfeeding and nipple pain were improved by frenotomy in 27 out of 28 neonates with ankyloglossia (Hogan et al. 2005). Messner et al. (2000) reported that 19% (8/36) of mothers had sore nipples without ankyloglossia, but there were no cases with difficulties in latching on to the breast.

15.2.2 Tongue Mobility and Speech

In a prospective study, 13 out of 14 patients with ankyloglossia noted symptoms including speech problems (50%) and physical limitations (57%), i.e., discomfort below the tongue, difficulties with kissing, licking the lips, eating an ice cream cone, keeping the teeth clean, and doing "tongue tricks" (Lalakea and Messner 2003). In the same study, the mobility of the tongue with and without ankyloglossia was measured in protruded and elevated tongue positions; the results were 15.5 ± 6.0 mm and 13.6 ± 8.0 mm with ankyloglossia and 32.0 ± 3.9 mm and 30.3 ± 4.9 mm without. The tongue functions were significantly improved after frenectomy. Speech problems can be induced by the restriction of tongue mobility by ankyloglossia. Interestingly, 7 of 14 patients mentioned above complained of speech difficulties, but their speech articulation was diagnosed as normal (Lalakea and Messner 2003). The relationship between ankyloglossia and speech is still controversial among speech pathologists. Some believe there is a potential association, while others find few or no correlations (Messner and Lalakea 2000). To our knowledge, no studies have compared the outcome of surgery with that of special training by speech pathologists without surgery.

15.2.3 Malocclusion and Gingival Recession

Ankyloglossia limits tongue mobility and makes the tongue adopt a low position, which causes forward and downward pressure on the mandible and can induce mandibular prognathism (Suter and Bornstein 2009). In an uncontrolled 38 case series study (age range 13–58 years) of frenectomy patients, skeletal class III occlusion was observed in 32 (Mukai et al. 1993). However, the relationship between ankyloglossia and jaw deformity remains unclear. Stoner and Mazdyasna reported that insertion of the LF in the interdental papilla could be relevant to lingual gingival recession (Stoner and Mazdyasna 1980), but this was not certain. For these reasons, there is no consensus on the indication for frenectomy. Dentists and guardians should seriously consider the advantages and disadvantages of surgery for ankyloglossia patients.

15.2.4 Surgical Techniques and Anatomy

In general, surgical techniques for ankyloglossia are classified into three different procedures: frenotomy, frenectomy, and frenoplasty. Newborns are operated under general anesthesia, but children able to have infiltration anesthesia to the tongue can be operated under local anesthesia. For uncooperative children, surgeries under general anesthesia could be preferable as the deep lingual vein and sublingual caruncle are at risk of injury from body movements. Frenotomy is the procedure for treating ankyloglossia with a simple incision, especially for newborns. This quick procedure can be selected in the dental office during the initial consultation. The newborns can breastfeed immediately after the operation. Frenotomy rarely causes bleeding as the LF is loose tissue with a poor blood supply (Hong et al. 2010; Masaitis and Kaempf 1996). A frenectomy is defined as complete excision of the LF. This technique is more invasive but more effective and is better performed before the patient develops

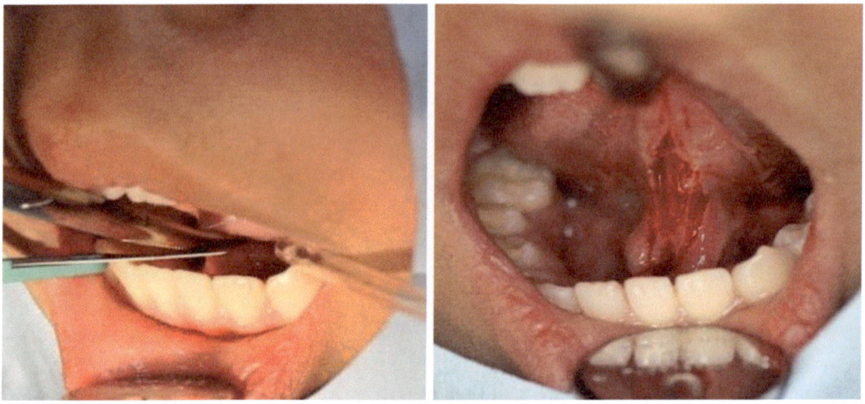

Fig. 15.4 Surgery for ankyloglossia

abnormal swallowing and speech (Manfro et al. 2010). Frenoplasty is the technique that incises genioglossus muscle to form the LF. The indications for frenectomy are patients without improvement of tongue mobility after frenotomy, patients with a thick LF, and patients whose ankyloglossia has recurred.

The anatomical structures that can be observed or injured during surgeries for ankyloglossia (Fig. 15.4) are from superficial to deep, the deep lingual vein, the terminal branch of the lingual nerve, and the deep lingual artery (Fig. 15.5). If the attachment of the LF is high in the alveolar mucosa, the sublingual caruncles and folds can be injured owing to the proximity of the incision line to the alveolar mucosa. However, complications following surgery for ankyloglossia are rarely reported. One study reported excessive bleeding in 3% of surgeries performed by otolaryngologists and 8% by pediatricians (Messner and Lalakea 2000).

15.3 Anterior Lingual Glands (Blandin-Nuhn Glands)

The anterior lingual glands, also known as the Blandin-Nuhn glands, are located in the ventral part of the tongue apex (Fig. 15.6). They are small minor salivary glands combining mucous and serous types, embedded within the musculature of the ventral side of the anterior tongue and covered by a thin layer of mucosa (Sugerman et al. 2000; Ellis et al. 1983). There are few anatomical studies of the anterior lingual glands. According to Ogawa and Mori (1956), they are oval with a long axis of 13.1 mm, a short axis of 5.3 mm, and a thickness of 5.4 mm. They are located in the inner quarter of the tongue apex, 5.2 mm behind it. Five to seven small ducts open into the ventral part of the tongue apex near the LF (Sugerman et al. 2000; Ellis et al. 1983). The intercalated ducts are composed of simple cuboidal epithelium and converge with the excretory ducts lined with columnar cells. Within the excretory ducts, there are regions of columnar cells with basal folds specialized for ion transport. These regions are reminiscent of striated ducts, and similar structures arise in the minor salivary glands of the lips (Tandler et al. 1994).

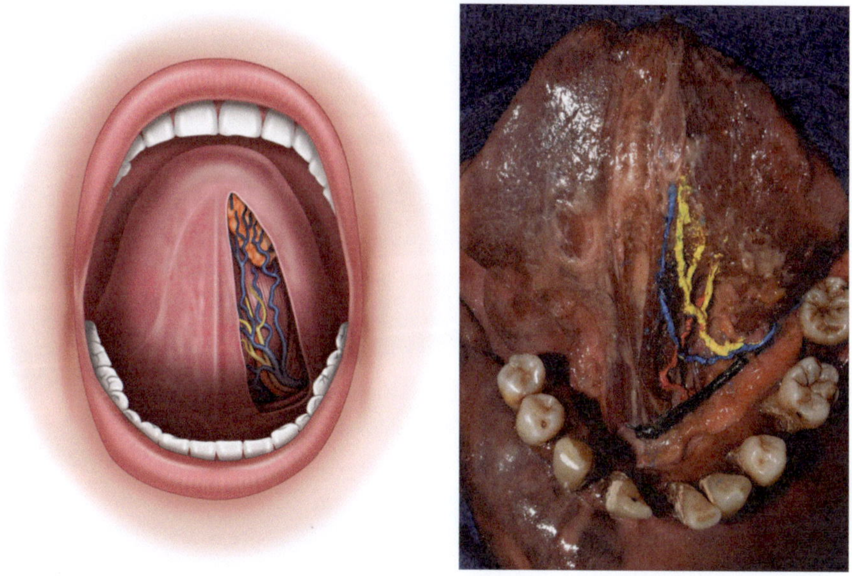

Fig. 15.5 The anatomical structures of the sublingual surface

Fig. 15.6 The positions of the anterior lingual glands

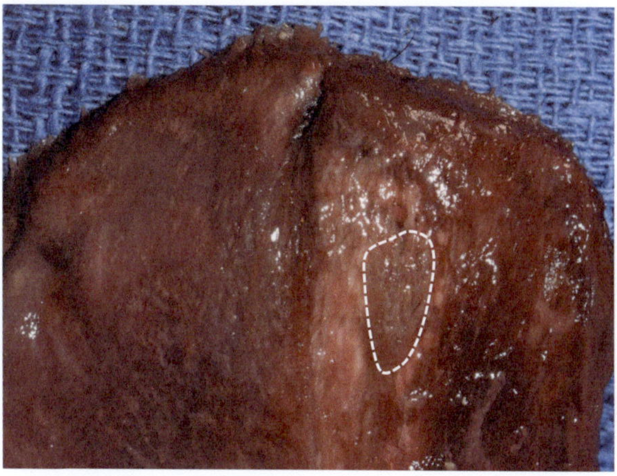

15.3.1 Mucoceles of the Blandin-Nuhn Glands

Mucoceles are common lesions induced by mechanical or physical injuries, which involve the rupture of a salivary duct and extravasation of the mucin into the surrounding soft tissues. The incidence of mucoceles in the Blandin-Nuhn glands has been reported as 2.3 to 15.4% (Harrison 1975; Jinbu et al. 2003; de Camargo Moraes et al. 2009; Saza et al. 1982) (Fig. 15.7).

The lesions are treated by excising the cystic mass including the causal glandular tissue and its associated duct. According to Baurmash 2003, large Blandin-Nuhn mucoceles are treated by completely unroofing the lesion along its entire periphery

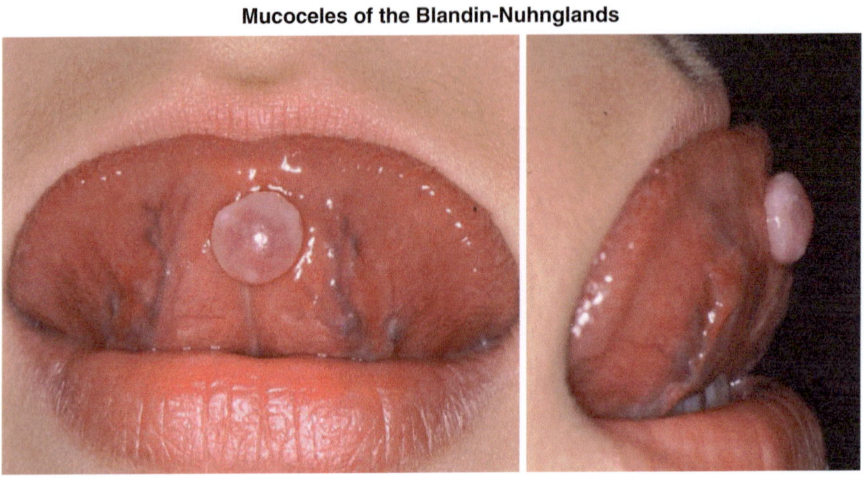

Fig. 15.7 Mucoceles of the Blandin-Nuhn glands

to visualize and remove all of the glands present. Small mucoceles are completely excised and closed. The superficial location and thinness of the wall of a mucocele demand extra care so that injury to the wall is precluded; it tends to rupture easily and this can result in recurrence (Guimaraes et al. 2006; Mandel and Kaynar 1992). Adachi et al. (2011) recommended that mucoceles of Blandin-Nuhn glands should be excised up to the muscle layer, including the small glands found in the surgical field, to preclude recurrence. Interestingly, Santos et al. (2012) reported an unusual case of large mucoceles of the Blandin-Nuhn glands, which appeared after 1 month from the lingual frenectomy. The authors assumed that suturing of the ventral surface of the mucosa could have included these glands, and this could have caused the mucocele.

The anatomical structures that can be observed or injured during excision of Blandin-Nuhn mucoceles are from superficial to deep, the deep lingual vein, inferior longitudinal muscle of the tongue, the anterior lingual glands, the lingual nerve, and the deep lingual artery. However, those complications have rarely been reported.

15.4 Deep Lingual Vein and Artery

The anatomy of the lingual vein is variable, but there are normally two large tributaries. The superiorly positioned dorsal lingual vein drains the dorsum of the tongue and accompanies the dorsal lingual branches of the lingual artery. The inferiorly positioned vein can be visualized along the ventral surface of the tongue. It merges with the sublingual vein and follows the hypoglossal nerve to become the lingual vein. Finally, the lingual vein merges with either the facial or the anterior division of the retromandibular vein to form the common facial vein, which drains into the internal jugular vein.

References

Adachi P, Soubhia AM, Horikawa FK et al (2011) Mucocele of the glands of Blandin–Nuhn—clinical, pathological, and therapeutical aspects. Oral Maxillofac Surg 15:11–13
Amir LH, James JP, Beatty J (2005) Review of tongue-tie release at a tertiary maternity hospital. J Paediatr Child Health 41:243–245
Baurmash HD (2003) Mucocele and ranulas. J Oral Maxillofac Surg 61:369–378
de Camargo Moraes P, Bönecker M, Furuse C et al (2009) Mucocele of the gland of Blandin–Nuhn: histological and clinical findings. Clin Oral Invest 13:351–353
Ellis E 3rd, Scott R, Upton LG (1983) An unusual complication after excision of a recurrent mucocele of the anterior lingual gland. Oral Surg Oral Med Oral Pathol 56:467–471
Friend GW, Harris EF, Mincer HH et al (1990) Oral anomalies in the neonate, by race and gender, in an urban setting. Pediatr Dent 12:157–161
García Pola MJ, González García M, García Martín JM et al (2002) A study of pathology associated with short lingual frenum. ASDC J Dent Child 69:59–62
Guimaraes MS, Hebling J, Filho VA et al (2006) Extravasation mucocele involving the ventral surface of the tongue (glands of Blandin-Nuhn). Int J Paediatr Dent 16:435–439
Harrison JD (1975) Salivary mucoceles. Oral Surg Oral Med Oral Pathol 39:268–278
Hazelbaker AK (1993) The Assessment Tool for Lingual Frenulum Function (ATLFF): use in a lactation consultant private practice. Pacific Oaks College; Thesis, Pasadena, CA
Heller J, Gabbay J, O'Hara C et al (2005) Improved ankyloglossia correction with four-flap Z-frenuloplasty. Ann Plast Surg 54:623–628
Hogan M, Westcott C, Griffiths M (2005) Randomized, controlled trial of division of tongue-tie in infants with feeding problems. J Paediatr Child Health 41:246–250
Hong P, Lago D, Seargeant J et al (2010) Defining ankyloglossia: a case series of anterior and posterior tongue ties. Int J Pediatr Otorhinolaryngol 74:1003–1006
Horton CE, Crawford HH, Adamson JE et al (1969) Tongue-tie. Cleft Palate J 6:8–23
Jinbu Y, Kusama M, Itoh H et al (2003) Mucocele of the glands of Blandin–Nuhn: clinical and histopathologic analysis of 26 cases. Oral Surg Oral Med Oral Pathol Oral Radiol Endod 95:467–470
Klockars T, Pitkaranta A (2009) Inheritance of ankyloglossia (tongue-tie). Clin Genet 75:98–99
Kotlow LA (1999) Ankyloglossia (tongue-tie): a diagnostic and treatment quandary. Quintessence Int 30:259–262
Lalakea ML, Messner AH (2003) Ankyloglossia: the adolescent and adult perspective. Otolaryngol Head Neck Surg 128:746–752
Mandel L, Kaynar A (1992) Mucocele of the gland of Blandin–Nuhn. NY State Dent J 58:40–41
Manfro AR, Manfro R, Bortoluzzi MC (2010) Surgical treatment of ankyloglossia in babies: case report. Int J Oral Maxillofac Surg 39:1130–1132
Marcano AC, Doudney K, Braybrook C et al (2004) TBX22 mutations are a frequent cause of cleft palate. J Med Genet 41:68–74
Masaitis NS, Kaempf JW (1996) Developing a frenotomy policy at one medical center: a case study approach. J Hum Lact 12:229–232
Messner AH, Lalakea ML (2000) Ankyloglossia: controversies in management. Int J Pediatr Otorhinolaryngol 54:123–131
Messner AH, Lalakea ML, Aby J et al (2000) Ankyloglossia: incidence and associated feeding difficulties. Arch Otolaryngol Head Neck Surg 126:36–39
Mukai S, Mukai C, Asaoka K (1993) Congenital ankyloglossia with deviation of the epiglottis and larynx: symptoms and respiratory function in adults. Ann Otol Rhinol Laryngol 102:620–624
Murakami M (1978) The alveolar mucous membrane where the frenulum of the tongue extends to. Jpn J Oral Biol 20:650–658
Ogawa K, Mori H (1956) Local anatomical study of the anterior lingual gland. Tokyo Dental College Anatomical Results Collection 1
Pauws E, Moore GE, Stanier P (2009) A functional haplotype variant in the TBX22 promoter is associated with cleft palate and ankyloglossia. J Med Genet 46:555–561

Ruffoli R, Giambelluca MA, Scavuzzo MC et al (2005) Ankyloglossia: a morphofunctional investigation in children. Oral Dis 11:170–174

Santos T de S, Filho PR, Piva MR et al (2012) Mucocele of the glands of Blandin-Nuhn after lingual frenectomy. J Craniofac Surg 23:e657–e658

Saza H, Shinohara M, Tomoyose Y et al (1982) Clinicostatistical study of salivary mucoceles. Jpn J Oral Maxillofac Surg 28:1545–1550

Schoenwolf GC, Bleyl S, Brauer PR et al (2015) Larsen's human embryology, 5th edn. Churchill Livingstone, Edinburgh

Segal LM, Stephenson R, Dawes M et al (2007) Prevalence, diagnosis, and treatment of ankyloglossia: methodologic review. Can Fam Physician 53:1027–1033

Stoner JE, Mazdyasna S (1980) Gingival recession in the lower incisor region of 15-year-old subjects. J Periodontol 51:74–76

Sugerman PB, Savage NW, Young WG (2000) Mucocele of the anterior lingual salivary glands (glands of Blandin and Nuhn): report of 5 cases. Oral Surg Oral Med Oral Pathol Oral Radiol Endod 90:478–482

Suter VG, Bornstein MM (2009) Ankyloglossia: facts and myths in diagnosis and treatment. J Periodontol 80:1204–1219

Tandler B, Pinkstaff CA, Riva A (1994) Ultrastructure and histochemistry of human anterior lingual salivary glands (glands of Blandin and Nuhn). Anat Rec 240:167–177

Tsaousoglou P, Topouzelis N, Vouros I, Sculean A (2015) Diagnosis and treatment of ankyloglossia: a narrative review and a report of three cases. Quintessence Int 47:523–534

Wallace H, Clarke S (2006) Tongue tie division in infants with breast feeding difficulties. Int J Pediatr Otorhinolaryngol 70:1257–1261

Part VI

Lip

Anatomy and Variations of the Labial Frena

16

Koichi Watanabe and Yoko Tabira

16.1 Introduction

The labial frenum is a fibrous band of tissue located in the middle of the oral vestibule and attached to the bone of the mandible or maxilla (Fig. 16.1). Frena can vary in number and development. Hyperplasia can occur in certain syndromes and in non-syndromic patients, leading to a median diastema or contraction of lip movement. The frenum can also be absent or hypoplastic in some cases. In this chapter we describe the anatomy and variations of the labial frenum.

16.2 Anatomy of Labial Frena

A frenum is a mucous membrane fold that attaches the lip or the cheek to the alveolar mucosa, gingiva, and underlying periosteum. Histological sections show that the labial frenum is composed of alveolar tissue, fibrous connective tissue, and a few striated muscle fibers (Noyes 1935). The striated muscle fibers arise from the bilateral muscle fibers of the lip either side of the midline, pass diagonally medially and posteriorly, and terminate before reaching the alveolar process. The connective tissue between the bilateral incisivus labii superioris muscles corresponds to the upper labial frenum in the maxilla (Iwanaga 2017a, b) (Fig. 16.2). Conversely, in the mandible, the connective tissue between the bilateral mentalis muscles corresponds with

K. Watanabe (✉) · Y. Tabira
Division of Gross and Clinical Anatomy, Department of Anatomy, Kurume University School of Medicine, Kurume, Japan
e-mail: yoko_tabira@med.kurume-u.ac.jp

J. Iwanaga, R. S. Tubbs (eds.), *Anatomical Variations in Clinical Dentistry*,
https://doi.org/10.1007/978-3-319-97961-8_16

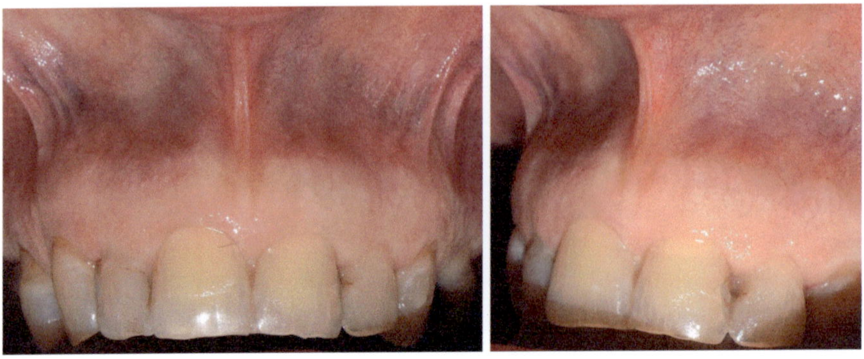

Fig. 16.1 Upper labial frenum

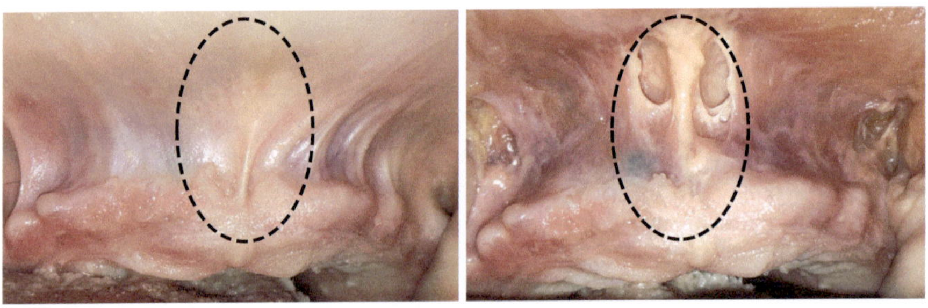

Fig. 16.2 Upper labial frenulum and underlining connective tissue

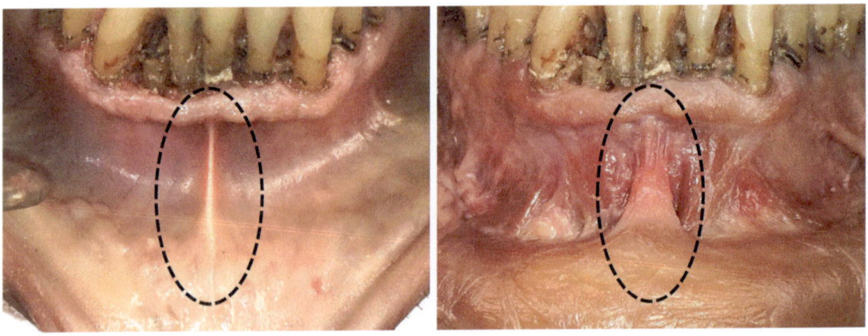

Fig. 16.3 Lower labial frenum and underlining connective tissue

the lower labial frenum (Fig. 16.3). Generally, the upper labial frenum can be observed more clearly than the lower one. The difference in appearance results from the characteristics of the connective tissue. The upper labial frenum has taut and thin connective tissue, whereas the lower one has taut but thick and flat connective tissue (Iwanaga 2017a, b).

16.3 Anatomical Variations of the Labial Frena

Both terms, *frenum* and *frenulum*, are used in the literature. Mintz defined a frenum as a mucosal attachment of a loose structure to a more rigid one and a frenulum as a small frenum (Mintz 2005). In most cases, the labial frenum attaches to the mucogingival junction. However, its development can vary. Some patients have well-developed frena extending beyond the interdental papilla, while others have normal frena inserting into the mucogingival junction. Frenectomies in the latter group of patients can preclude unfavorable outcomes from orthodontic treatment, such as the development of a median midline diastema. However, it is still controversial whether frenectomy contributes to gingival recession and peri-implant diseases in the midline area of the maxilla (Delli et al. 2013). Mirko et al. (1974) advocated the classification of frenum development into four types and reported the frequency of those four types in the upper and lower lips as follows (Fig. 16.4).

16.3.1 Mucosal Attachment

The frenum attaches to the mucogingival junction. This is the most common attachment, with an incidence of 46.5% in the upper lip and 92.1% in the lower.

16.3.2 Gingival Attachment

The frenum attaches to the gingiva. This is the second most common attachment, with an incidence of 34.3% in the upper lip and 6.5% in the lower.

16.3.3 Papillary Attachment

The frenum attaches to the interdental papilla. This attachment is less common, with an incidence of 3.1% in the upper lip and 0.2% in the lower.

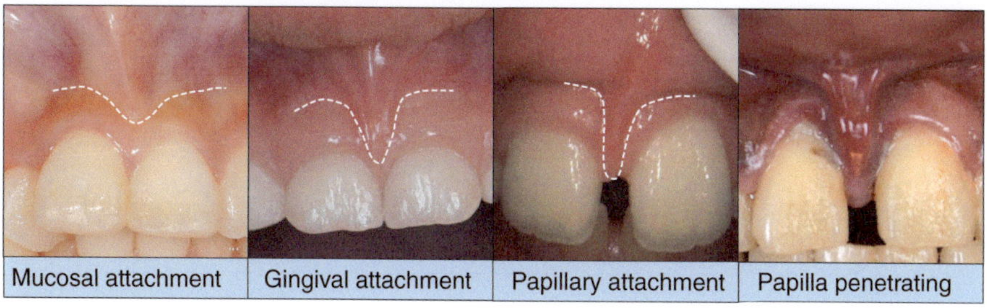

Fig. 16.4 Various attachments of the labial frenum (Courtesy of Dr. Koshi Shiromoto, White Dental Office Hanzomon)

16.3.4 Papilla Penetrating

The frenum attaches to the gingiva and passes through the interdental papilla to the palate or floor of the mouth. This is the third most common attachment, occurring with an incidence of 16.1% in the upper lip and 1.2% in the lower.

Frena, especially upper labial frena, sometimes develop an unclassifiable shape with accompanying nodules or appendices (Fig. 16.5). Mohan et al. (2014) modified Sewerin's classification system (1971) and proposed a new one:

1. Simple frenum (simple single, simple double, simple triple); 66.21% incidence
2. Frenum with nodule (at labial 1/3, at middle 1/3, at alveolar 1/3, multiple nodules); 19.92% incidence
3. Frenum with appendix (at labial 1/3, at middle 1/3, at alveolar 1/3); 6.38% incidence
4. Tectolabial frenum (simple tectolabial, tectolabial with lateral appendix, tectolabial with nodule); 5.63% incidence
5. New types (inverted Y frenum, frenum with reces, bifid frenum, trifid frenum)

The highest incidence was 63.79%, the simple single frenum being the most common type. However, this indicates that approximately 35% of cases have anatomical variations.

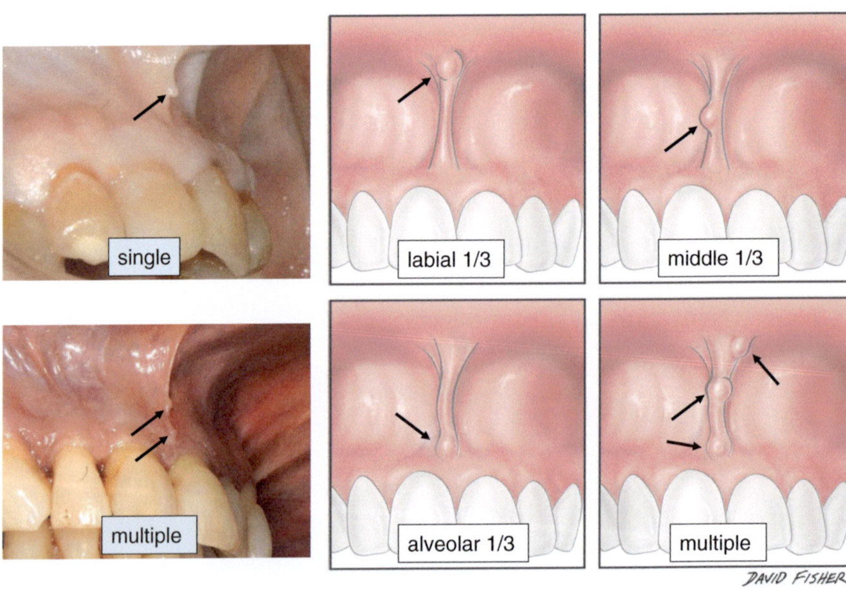

Fig. 16.5 Frenum with nodules

16.4 Diseases Resulting in Absence of the Labial Frenum

It is well-known that absence of the labial frenum is part of the profile of certain diseases including Ehlers–Danlos syndrome, infantile hypertrophic pyloric stenosis, and holoprosencephaly (Mintz 2005). We describe the details of these syndromes in this section.

16.4.1 Ehlers–Danlos Syndrome

Ehlers–Danlos syndrome (EDS) is an inheritable connective tissue disorder characterized by joint hypermobility, skin hyperextensibility, and tissue fragility. This syndrome is believed to be caused by mutation of a gene encoding a specific collagen type or an enzyme involved in collagen maturation. The International Ehlers–Danlos Syndrome Consortium proposed a revised EDS classification in 2017, which recognizes 13 subtypes (Malfait 2017). The incidence of this syndrome, including every subtype, is one in 5000 individuals. De Felice (2001) examined 12 consecutive patients with EDS and 154 controls with no history of either inheritable connective tissue disorders or infantile hypertrophic pyloric stenosis, which also causes absence of the labial frenum. The results showed that absence of the inferior labial frenum has 100% sensitivity and 99.4% specificity (95% CI 96.4–99.9%) for identifying EDS, while absence of the superior lingual frenum has 71.4% sensitivity (95% CI 29.3–95.5%) and 100% specificity. In contrast, Böhm (2001) analyzed 16 EDS patients; nine had classic autosomal dominant EDS and seven had hypermobile EDS. The inferior labial and lingual frena were missing in one of the nine patients with classic EDS, but the superior labial frenum was present. Among the patients with hypermobile EDS, only two lacked the inferior labial frena.

16.4.2 Infantile Hypertrophic Pyloric Stenosis

Infantile hypertrophic pyloric stenosis (IHPS) causes abnormal thickening of the pyloric portion of the stomach, resulting in narrowing and elongation of the pyloric channel. Most infants with IHPS show forceful or projectile non-bilious vomiting after feeding during the first 2–12 weeks of life. The incidence of this disease ranges from 0.5 to 4.21 in 1000 live births, and the condition appears to be caused by the interaction of genetic and environmental factors. IHPS is more prevalent among first-born infants and has a 4–5 times greater prevalence in boys than girls (MacMahon 2006; Ranells 2011).

De Felice (2000) reported that 92% of 25 consecutive patients with IHPS but only 1.6% of 319 consecutive non-IHPS had a hypoplastic or absent lower labial frenum ($P < 0.001$).

16.4.3 Holoprosencephaly

Holoprosencephaly is a genetically and phenotypically heterogenous disorder involving the development of the forebrain and midface. The incidence is one in 16,000 live births. This disorder manifests with unusual facial features. Anophthalmia or cyclopia is evident, and in the most extreme cases, the mature nose is congenitally absent. Ocular hypertelorism and defects of the upper lip and/or nose are also observed in less severe cases (Belloni et al. 1996). Martin and Jones (1998) stated that the upper labial frenum was absent in 88% of 17 consecutive individuals with holoprosencephaly regardless of the severity of holoprosencephaly or other associated craniofacial defects.

16.5 Diseases Resulting in Hyperplasia of the Labial Frenum

The oral anomalies in some conditions, including Ellis–van Creveld syndrome, oral-facial-digital syndrome type 1, Pallister–Hall syndrome, and Opitz C syndrome, are reported to include hyperplasia of the labial frenum.

16.5.1 Ellis–van Creveld Syndrome

Ellis–van Creveld syndrome is a rare genetic disorder characterized by short-limb dwarfism, polydactyly, abnormal development of the fingernails, and congenital heart disease. This syndrome is also called chondroectodermal dysplasia. Its incidence is approximately seven in one million births. There is hypertrophy of the labial frenum, which is often manifold and broad with abnormal attachment. Other oral manifestations include submucosal clefts, absence of the vestibular sulcus, dystrophic philtrum, hypodontia, enamel hypoplasia, delayed eruption, teeth with abnormal forms (microdontia, conical teeth, dens in dente, and taurodontism), malocclusion, and diastema (Hattab et al. 1998; Hanemann et al. 2010).

16.5.2 Oral-Facial-Digital Syndrome Type 1 (Papillon–Leage–Psaume Syndrome)

Oral-facial-digital syndrome type 1 is a rare genetic disorder characterized by anomalies of the face, oral cavity, and digits. There are also anomalies in other areas including the central nervous system and kidneys. This syndrome results from an X-linked dominant allele, and the estimated incidence is 1 in 50,000–250,000 live births. The oral frena are hyperplastic. Other oral abnormalities in this syndrome include median pseudo-cleft of the upper lip, cleft palate, hamartoma of the tongue, and bifid tongue (Gurrieri et al. 2007).

16.5.3 Pallister–Hall Syndrome

Pallister–Hall syndrome is a genetic disorder characterized by a spectrum of anomalies including polydactyly, asymptomatic bifid epiglottis, and hypothalamic hamartoma. There are abnormal supernumerary frena extending from the buccal mucosa to the alveolar ridge. Other orofacial abnormalities can include micrognathia, microglossia, short midface, flat nasal bridge, and anteverted nostrils (Priyanka et al. 2013). Pallister–Hall syndrome patients sometimes express features of oral-facial-digital syndrome, including polydactyly and oral anomalies. This condition is called oral-facial-digital overlap syndrome (Biesecker 2017).

16.5.4 Opitz C Syndrome

The main feature of this syndrome is trigonocephaly, a congenital cranial anomaly resulting in a narrow pointed forehead. There are frenal abnormalities similar to Pallister–Hall syndrome (Priyanka et al. 2013), multiple attached buccal frena along with an unusual facial appearance and wide alveolar ridges (Mintz 2005). This syndrome can also involve abnormalities of facial appearance, limbs, and neck (Sargent et al. 1985).

16.6 Other Anatomical Variations of the Labial Frena

Mintz (2005) noted that patients with W syndrome, which is characterized by median cleft lip, mental retardation, and pugilistic facies, have a bifid upper labial frenum that causes notching of the midline in the upper lip. Median cleft lip is generally classed as either true or false. Ichida et al. (2009) stated that a true median cleft lip sometimes has two upper labial frena, whereas a false one has no upper labial frenum.

References

Belloni E, Muenke M, Roessler E et al (1996) Identification of sonic hedgehog as a candidate gene responsible for holoprosencephaly. Nat Genet 14:353–356

Biesecker LG. Pallister-Hall Syndrome GeneReviews® [Internet]. https://www.ncbi.nlm.nih.gov/books/NBK1465/ (Initial Posting: May 25, 2000; Last Update: May 18, 2017)

Böhm S, Martinez-Schramm AM, Beherens P (2001) Missing inferior labial and lingual frenula in Ehlers-Danlos syndrome. Lancet 358:1647

De Felice C, Di Maggio G, Zagordo L et al (2000) Hypoplastic or absent mandibular frenulum: a new predictive sign of infantile hypertrophic pyloric stenosis. J Pediatr 136:408–410

De Felice C, Toti P, Di Maggio G et al (2001) Absence of the inferior labial and lingual frenula in Ehlers-Danlos syndrome. Lancet 357:1500–1502

Delli K, Livas C, Sculean A et al (2013) Facts and myths regarding the maxillary midline frenum and its treatment: a systematic review of the literature. Quintessence Int 44:177–187

Gurrieri F, Franco B, Toriello H et al (2007) Oral–facial–digital syndromes: review and diagnostic guidelines. Am J Med Genet A 143:3314–3323

Hanemann JA, de Carvalho BC, Franco EC (2010) Oral manifestations in Ellis-van Creveld syndrome: report of a case and review of the literature. J Oral Maxillofac Surg 68:456–460

Hattab FN, Yassin OM, Sasa IS (1998) Oral manifestations of Ellis-van Creveld syndrome: report of two siblings with unusual dental anomalies. J Clin Pediatr Dent 22:159–165

Ichida M, Komuro Y, Yanai A (2009) Consideration of median cleft lip with frenulum labii superior. J Craniofac Surg 20:1370–1374

Iwanaga J, Takeuchi N, Oskouian RJ et al (2017a) Clinical anatomy of the frenulum of the oral vestibule. Cureus 9:e1410

Iwanaga J, Watanabe K, Schmidt CK et al (2017b) Anatomical study and comprehensive review of the incisivus Labii Superioris muscle: application to lip and cosmetic surgery. Cureus 9:e1689

MacMahon B (2006) The continuing enigma of pyloric stenosis of infancy: a review. Epidemiology 17:195–201

Malfait F, Francomano C, Byers P et al (2017) The 2017 international classification of the Ehlers–Danlos syndromes. Am J Med Genet C Semin Med Genet 175(1):8–26

Martin RA, Jones KL (1998) Absence of the superior labial frenulum in holoprosencephaly: a new diagnostic sign. J Pediatr 133:151–153

Mintz SM, Siegel MA, Seider PJ (2005) An overview of oral frena and their association with multiple syndromic and nonsyndromic conditions. Oral Surg Oral Med Oral Pathol Oral Radiol Endod 99:321–324

Mirko P, Miroslav S, Lubor M (1974) Significance of the labial frenum attachment in periodontal disease in man. Part I. Classification and epidemiology of the labial frenum attachment. J Periodontol 45:891–894

Mohan R, Soni PK, Krishna MK et al (2014) Proposed classification of medial maxillary labial frenum based on morphology. Dent Hypoth 5:16–20

Noyes HJ (1935) The anatomy of the Frenum Labii in new born infants. Angle Orthod 5:3–8

Priyanka M, Sruthi R, Ramakrishnan T et al (2013) An overview of frenal attachments. J Ind Soc Periodontol 17:12–15

Ranells JD, Carver JD, Kirby RS (2011) Infantile hypertrophic pyloric stenosis: epidemiology, genetics, and clinical update. Adv Pediatr 58:195–206

Sargent C, Burn J, Maraister M et al (1985) Trigonocephaly and the Opitz C syndrome. J Med Genet 22:39–45

Sewerin IB (1971) Prevalence of variations and anomalies of the upper labial frenum. Acta Odontol Scand 29:487–496

17 Anatomy and Variations of the Lip

Koichi Watanabe and Tsuyoshi Saga

17.1 Introduction

The lips are two flaps surrounding the oral orifice. As they are located at the entrance to the digestive tract, the external surface is covered with skin and the internal surface with mucosa. Externally, the appearance of the lip can vary because of syndromes affecting the face. Lip shape is usually defined by the shape of the philtrum. The lateral border of the lip and the nasolabial fold differ among individuals in shape and length. Internally, the labial salivary glands can be affected by disease and aging. The muscles in the lips can also be altered by certain diseases. In this chapter, we describe the anatomy and variations of the lip.

17.2 Anatomy of the Lip (Fig. 17.1)

The upper lip and the lower lip surround the oral orifice, the entrance to the oral cavity. The location at which the upper and lower lips meet laterally is known as the oral commissure. The lips are covered with skin externally and mucosa internally. The external surface of the free edge of each lip is red in color, and this zone is called the vermillion or rubor labiorum.

The skin-covered portion of the lips contains hair follicles, sebaceous glands, and sweat glands. It lies adjacent to other facial structures: the lateral border with the cheek forms the nasolabial fold, and the caudal border with the mentum forms the labiomental fold. These two folds are poorly defined in the young but become more apparent with age. There is a shallow vertical groove in the midline of the

K. Watanabe (✉) · T. Saga
Kurume University School of Medicine, Kurume, Fukuoka, Japan
e-mail: saga@med.kurume-u.ac.jp

J. Iwanaga, R. S. Tubbs (eds.), *Anatomical Variations in Clinical Dentistry*,
https://doi.org/10.1007/978-3-319-97961-8_17

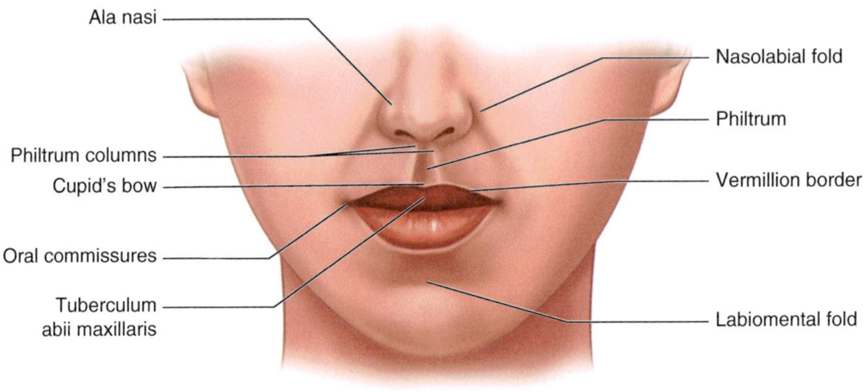

Fig. 17.1 Anatomy of the lip

skin-covered portion of the upper lip known as the philtrum, which connects the base of the columella to the vermillion border. The bilateral borders of the philtrum are demarcated by convex ridge lines called the philtrum columns.

In contrast to the skin and mucosa of the lips, the lip vermilion contains no hair follicles, sebaceous glands, sweat glands, or salivary glands. The tubercle is located in the center of the upper lip. The border between the lip vermilion and skin is known as the vermilion border. It is sometimes also called the white skin roll because of its slightly convex shape, similar to a ridge line. In the middle of the upper lip, the vermilion border forms a double curved shape called the Cupid's bow.

The lip mucosa contains minor salivary glands known as the labial glands. The border between the lip mucosa and the vermillion can be clearly seen as a demarcation between wet and dry mucosa. It is called the wet-dry line.

The muscles of the lips are sphincter and dilator muscles. The sphincter is made up of orbicularis oris muscle, which surrounds the oral orifice inside the lip. The dilator muscles include levator labii superioris, zygomaticus minor and major, risorius, incisivus labii superioris, incisivus labii inferioris, levator anguli oris, depressor anguli oris, depressor labii inferioris, mentalis, and buccinator, which are arranged radially and attach to the orbicularis oris, the muscle fibers becoming intertwined (Iwanaga et al 2017a; Iwanaga et al 2017b; Ethunandan 2016).

17.3 Anatomical Variations in the Skin of the Lip

17.3.1 Philtrum

Syndromes entailing specific facial appearances often result in abnormal length, width, and/or depth of the philtrum. The philtrum is located in the midline of the upper lip and connects the columella and vermillion. A deformity in it often alters the shape of the upper lip relative to its surrounding structures. Possible causes of philtrum abnormality include exposure to certain drugs during pregnancy and chromosomal aberrations, but in some syndromes, the etiology is unknown. We describe syndromes associated with altered philtrum development in this subchapter.

17.3.1.1 Aarskog Syndrome (Long Philtrum)

Aarskog syndrome (facio-digital-genital syndrome) is an X-linked disorder characterized by short stature, macrocephaly, and facial, genital, and skeletal anomalies (Berman et al. 1975; Schwartz et al. 2000). In this syndrome, the upper lip has a long philtrum and thin vermillion border (Berman et al. 1975).

17.3.1.2 Cohen Syndrome (Short Philtrum)

Cohen syndrome is associated with facial dysmorphism, developmental delay, and visual disability. The philtrum is short and upturned and the upper lip is thin (Chandler et al. 2003).

17.3.1.3 DiGeorge Syndrome (Short Philtrum)

DiGeorge syndrome comprises thymic hypoplasia, hypocalcemia, outflow tract defects of the heart, and dysmorphic facies. The philtrum is often short in this syndrome (Wilson et al. 1993).

17.3.1.4 Fetal Alcohol Syndrome (Smooth Philtrum)

Fetal alcohol syndrome is considered part of fetal alcohol spectrum disorder. It is caused by prenatal alcohol exposure, and affected patients display characteristic facial dysmorphology, growth restriction, and central nervous system/neurodevelopmental abnormalities. The upper lip is thin with a long and smooth philtrum (Sokol et al. 2003).

17.3.1.5 Floating–Harbor Syndrome (Short and Smooth Philtrum)

Floating–Harbor syndrome is characterized by short stature, delayed osseous maturation, expressive language deficits, and a distinctive facial appearance. The upper lip has a short philtrum, and the mouth is wide with a thin vermillion border (Hood et al. 2012).

17.3.1.6 Pallister–Killian Syndrome

Pallister–Killian syndrome is caused by tetrasomy 12p mosaicism. The characteristic features are profound motor deficit and mental retardation (Horn et al. 1995). Patients with this condition have a long and simple philtrum with a prominent upper lip (Schinzel 1991).

17.3.1.7 Trichorhinophalangeal Syndrome (Long and Flat Philtrum)

Trichorhinophalangeal syndrome is an autosomal dominant condition characterized by craniofacial and skeletal abnormalities. Patients with this syndrome have a long and flat philtrum and a thin upper lip vermillion border (Momeni et al. 2000).

17.3.1.8 Valproate Embryopathy (Long and Flat Philtrum)

Valproate is used to treat epilepsy and bipolar disorder. Exposure to it during pregnancy causes the anomalies seen in fetal valproate embryopathy, which is characterized by a distinctive facial appearance and developmental delay. The upper lip has a long philtrum and thin vermillion border (Rodríguez-Vázquez et al. 2007).

17.3.1.9 Weaver Syndrome

Weaver syndrome is characterized by excessive growth of prenatal onset, macrocephaly, accelerated skeletal maturation, and delayed psychomotor development (Meinecke et al. 1983). Patients with this syndrome have a characteristic facial appearance including a broad and prominent forehead, hypertelorism, and a long and accentuated philtrum (Majewski 1981).

17.3.2 Nasolabial Fold

The nasolabial fold is located at the border between the lip and the cheek, at the junction of the ala nasi, cheek, and upper lip at the most superior point, and it curves downward and laterally to be lost below and lateral to the corner of the mouth (Rubin et al. 1989). Generally, the direction of the fold is classified into three types: straight, convex, and concave (Rubin et al. 1989; Zufferey 1992).

The straight type is almost straight from the tip of the ala to the corner of the mouth. The convex type is curved and convex laterally, and the concave type is curved and concave laterally.

Pessa et al. (1998) modified this three-type classification by considering the length of the nasolabial fold. The fold was classed as either short, extended, or continuous, depending on whether its caudal termination was above, at, or below the corner of the mouth, respectively. The nasolabial fold was convex, straight, and concave in 60%, 30%, and 10% of cases, respectively, and its length was short, extended, and continuous in 38%, 42%, and 20% of cases, respectively.

The depth of the nasolabial fold changes with age. Zufferey (1992) found that the fold is absent in children, initially appearing after extensive facial expression at around 25 years old and becoming prominent at 35 years old. It disappears in patients with facial nerve palsy.

17.3.3 Other Anatomical Variations of the Skin of the Lip

Various degrees of cleft lip sometimes occur and can be unilateral (right or left side) or bilateral, and complete or incomplete. In complete cleft lip, the cleft is located in the upper lip between the medial and lateral lip from the nasal base to the upper lip border. Cleft lip sometimes accompanies various degrees of cleft palate. Median cleft lip is a rare anomaly, a vertical cleft in the center of the upper lip. Its incidence is about 1 in 1000,000 in the general population (Koh 2016). An oblique facial cleft (Tessier Nos. 3, 4, and 5) can also affect the upper lip. This cleft connects the upper lip to the lower eyelid or the alveolar process of the maxilla and the orbit (Tessier 1976; Stretch and Poole 1990). A transverse facial cleft can occur at the oral commissure, when it is also called macrostomia or Tessier No. 7 cleft. The incidence of this anomaly is about 1 in 80,000 live births (Tessier 1976; Rogers 2007).

17.4 Anatomical Variations in the Lip Mucosa

17.4.1 Labial Glands

Labial glands are located in the mucosal surface of the lips and secrete both mucous and serous saliva (Gleeson 2016). Most of them lie in the connective tissue of the lamina propria, but some are located in the muscle layer. Magnetic resonance imaging has shown that these glands bilaterally span the first premolars in the upper lip and the second premolars in the lower lip (Sumi 2007).

The glands located in the posterior parts of both the upper and lower lips, spanning from the canine to premolar region, are thicker than those in the anterior (incisor) parts (Sumi 2007). There are more glands in the anterior than the posterior part (Fig. 17.2). The labial glands stratify in 1–2 layer clusters of discrete gland areas in the upper lip and 1–3 layers in the lower lip (Sumi 2007).

Sjögren's syndrome is an autoimmune disease that causes sicca symptoms including dry mouth and eyes. This suggests that the lacrimal and salivary glands are most affected by the disease, which is often diagnosed by a labial gland biopsy (Kassan and Moutsopoulos 2004). The gland areas, particularly in the upper lip, become smaller in patients with Sjögren's syndrome (Sumi 2007).

A cyst, mucocele, or mucous cyst sometimes develops on the mucosal aspect of the lip. Mucoceles result from the rupture of a salivary gland duct and extravasation of mucus into the surrounding soft tissue. This cyst usually appears on the lower lip

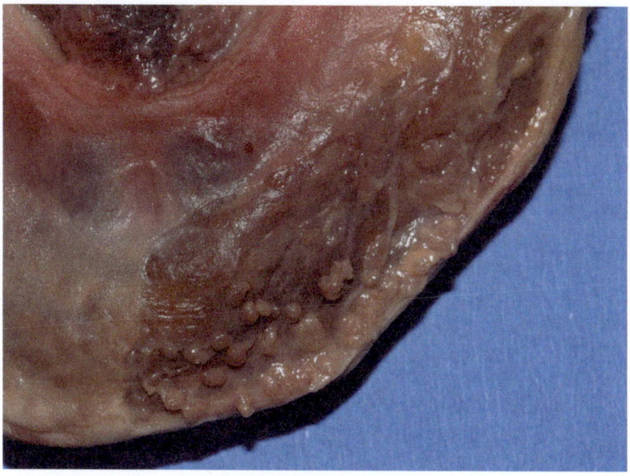

Fig. 17.2 Minor salivary glands in the lower lip

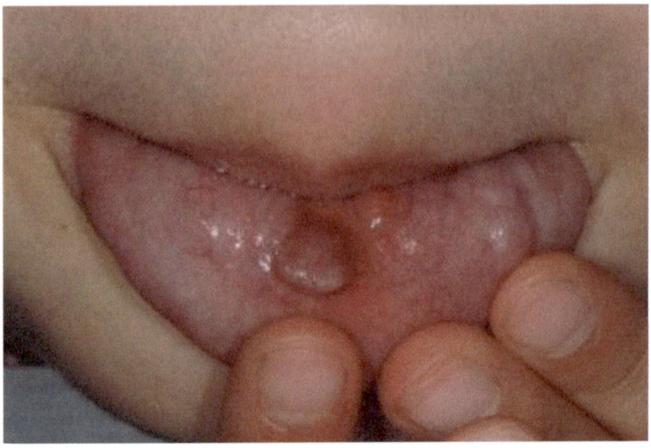

Fig. 17.3 Mucocele on the lower lip

(Fig. 17.3); cysts on the upper lip are very rare (Chi et al. 2011). Salivary duct cysts similar to mucoceles can also occur, lined with salivary duct epithelium. They are less common and tend to appear more often in older populations, in contrast to mucoceles (Chi et al. 2011).

Aging also results in histological changes in the labial glands. Drummond and Chisholm (1984) found that the fibrous tissue content of labial glands increases with aging. Scott (1980) showed that acinar atrophy, duct dilation, and hyperplasia of the labial glands also increase with aging.

17.4.2 Other Anomalies of the Lip Mucosa

Rarely, excessive or redundant hypertrophic tissue of the lip mucosa protrudes downward and forms a double lip. This is more common on the upper lip but can occur on the lower one. It can be congenital or acquired or occur among the symptoms of Ascher's syndrome (Martins et al. 2004; Daniels 2010).

17.5 Muscles

17.5.1 Orbicularis Oris

Cleft lip causes failure of continuity of orbicularis oris muscle on the upper lip. Orbicularis oris loses continuity at the cleft and terminates beneath the alar base laterally and beneath the base of the columella medially. In bilateral cleft lip, the muscle terminates beneath the bilateral base of the ala, and the medial lip, or prolabium, is composed of connective tissue (Fara 1990).

17.5.2 Depressor Anguli Oris

Congenital hypoplasia of depressor anguli oris has been reported (Gupta 2009). This causes facial asymmetry while crying. The incidence of this anomaly is 3–6 in 1000 live births. It is usually associated with cardiac, gastrointestinal, and genitourinary anomalies (Lin et al. 1997; Caksen et al. 2004).

References

Berman P, Desjardins C, Fraser FC (1975) The inheritance of the Aarskog facial-digital-genital syndrome. J Pediatr 86:885–891

Caksen H, Odabas D, Tuncer O (2004) A review of 35 cases of asymmetric crying facies. Genet Couns 15:159–165

Chandler KE, Kidd A, Al-Gazali L et al (2003) Diagnostic criteria, clinical characteristics, and natural history of Cohen syndrome. J Med Genet 40:233–241

Chi AC, Lambert PR, Richardson MS et al (2011) Oral mucoceles: a clinicopathologic review of 1,824 cases, including unusual variants. J Oral Maxillofac Surg 69:1086–1093

Daniels JSM (2010) Congenital double upper lip: a case report and review of the literature. Saudi Dent J 22:101–106

Drummond JR, Chisholm DM (1984) A qualitative and quantitative study of the ageing human labial salivary glands. Arch Oral Biol 29:151–155

Ethunandan MG (2016) Chapter 9: Oral cavity. In: Brennan PA, Mahadevan V, Evans BT (eds) Clinical head and neck anatomy for surgeons. CRC Press, Boca Raton, pp 77–80

Fara M (1990) Chapter 51. The musculature of cleft lip and palate. In: JG MC (ed) Plastic surgery general principles, vol 1. WB Saunders Co, Philadelphia, pp 2598–2612

Gleeson M (2016) Chapter 31. Oral cavity. In: Standring S (ed) Gray's anatomy, 41st edn. Elsevier, Amsterdam, pp 507–508

Gupta R, Prasad P (2009) Congenital hypoplasia of depressor angularis oris muscle. Med J Armed Forces India 65:188–189

Hood RL, Lines MA, Nikkel SM et al (2012) Mutations in SRCAP, encoding SNF2-related CREBBP activator protein, cause Floating-Harbor syndrome. Am J Hum Genet 90:308–313

Horn D, Majewski F, Hildebrandt B et al (1995) Pallister-Killian syndrome: normal karyotype in prenatal chorionic villi, in postnatal lymphocytes, and in slowly growing epidermal cells, but mosaic tetrasomy 12p in skin fibroblasts. J Med Genet 32:68–71

Iwanaga J, Watanabe K, Schmidt CK et al (2017a) Anatomical study and comprehensive review of the incisivus labii superioris muscle: application to lip and cosmetic surgery. Cureus 15:e1689

Iwanaga J, He P, Watanabe K et al (2017b) Intraoral observation of the mentalis and incisivus labii inferioris muscles. J Craniofac Surg 28:2159–2161

Kassan SS, Moutsopoulos HM (2004) Clinical manifestations and early diagnosis of Sjögren syndrome. Arch Intern Med 164:1275–1284

Koh KS, Kim DY, Oh TS (2016) Clinical features and management of a median cleft lip. Arch Plast Surg 43:242–247

Lin DS, Huang FY, Lin SP (1997) Frequency of associated anomalies in congenital hypoplasia of depressor anguli oris muscle: a study of 50 patients. Am J Med Genet 71:215–218

Majewski F, Ranke M, Kemperdick H, Schmidt E (1981) The Weaver syndrome: a rare type of primordial overgrowth. Eur J Pediatr 137:277–282

Martins WD, Westphalen FH, Sandrin R et al (2004) Congenital maxillary double lip: review of the literature and report of a case. J Can Dent Assoc 70:466–468

Meinecke P, Schaefer E, Engelbrecht R (1983) The Weaver syndrome in a girl. Eur J Pediatr 141:58–59
Momeni P, Glöckner G, Schmidt O et al (2000) Mutations in a new gene, encoding a zinc-finger protein, cause tricho-rhino-phalangeal syndrome type I. Nat Genet 24:71–74
Pessa JE, Zadoo VP, Adrian EK Jr et al (1998) Variability of the midfacial muscles: analysis of 50 hemifacial cadaver dissections. Plast Reconstr Surg 102:1888–1893
Rodríguez-Vázquez M, Carrascosa-Romero MC et al (2007) Congenital gingival hyperplasia in a neonate with foetal valproate syndrome. Neuropediatrics 38:251–252
Rogers GF, Mulliken JB (2007) Repair of transverse facial cleft in hemifacial microsomia: long-term anthropometric evaluation of commissural symmetry. Plast Reconstr Surg 120:728–737
Rubin LR, Mishriki Y, Lee G (1989) Anatomy of the nasolabial fold: the keystone of the smiling mechanism. Plast Reconstr Surg 83:1–8
Schinzel A (1991) Tetrasomy 12p (Pallister-Killian syndrome). J Med Genet 28:122
Schwartz CE, Gillessen-Kaesbach G, May M et al (2000) Two novel mutations confirm FGD1 is responsible for the Aarskog syndrome. Eur J Hum Genet 8:869
Scott J (1980) Qualitative and quantitative observations on the histology of human labial salivary glands obtained post mortem. J Biol Buccale 8:187–200
Sokol RJ, Delaney-Black V, Nordstrom B (2003) Fetal alcohol spectrum disorder. JAMA 290:2996–2999
Stretch JR, Poole MD (1990) Nasolacrimal abnormalities in oblique facial clefts. Br J Plast Surg 43:463–467
Sumi M, Yamada T, Takagi Y, Nakamura T (2007) MR imaging of labial glands. AJNR Am J Neuroradiol 28:1552–1556
Tessier P (1976) Anatomical classification of facial, cranio-facial and latero-facial clefts. J Maxillofac Surg 4:69–92
William R (2015) Chapter 2.13. Abnormalities of the Philtrum. In: The bedside dysmorphologist. Oxford University Press, Oxford, pp 48–51
Wilson DI, Burn J, Scambler P et al (1993) DiGeorge syndrome: part of CATCH 22. J Med Genet 30:852–856
Zufferey J (1992) Anatomic variations of the nasolabial fold. Plast Reconstr Surg 89:225–231

Part VII

Temporomandibular Joint

Anatomy and Variations of the Temporomandibular Joint

18

Rebecca C. Ramdhan and Joe Iwanaga

18.1 Introduction

The temporomandibular joint (TMJ) is classified as a synovial joint that permits gliding, rotation, elevation, and depression and normally functions in mastication, suckling, swallowing, yawning, speaking, and biting. The mandibular fossa, the articular eminence of the temporal bone, and the mandibular condyle all contribute to the TMJ (Fig. 18.1). This joint can be indicated by the eminentia mandibularis, a bony protuberance on the floor of the middle cranial fossa (Tubbs et al. 2008). The articular eminence is a convex bony elevation on the root of the zygomatic process that characterizes the most anterior boundary of the mandibular fossa. Both the fibrous capsule and the lateral ligament (Fig. 18.2) attach to the articular eminence, and the capsule also attaches to the articular cartilage of the temporal bone and the neck of the mandible. The articular surfaces of the TMJ are uniquely lined with fibrocartilage rather than the normal hyaline cartilage. Its joint cavity is divided into two (superior and inferior joint spaces) by an articular disc (Fig. 18.3), and this disc attaches to the capsule separating the two joint spaces lined with synovial membranes. The synovial membranes do not extend to cover the disc or the articular surfaces. The joint moves and functions via the lateral, stylomandibular, and sphenomandibular ligaments. The muscles that allow for TMJ articulation are the temporalis, masseter, medial, and lateral pterygoids, collectively known as the muscles of mastication. The suprahyoid muscle group except the stylohyoid (digastric, mylohyoid, and geniohyoid) is also responsible for mandibular movement. The synovial joints are supplied by sensory nerve endings (mainly proprioceptive) with pain and stretch receptors. The articular capsules and ligaments are highly

R. C. Ramdhan · J. Iwanaga (✉)
Seattle Science Foundation, Seattle, WA, USA
e-mail: joei@seattlesciencefoundation.org

J. Iwanaga, R. S. Tubbs (eds.), *Anatomical Variations in Clinical Dentistry*,
https://doi.org/10.1007/978-3-319-97961-8_18

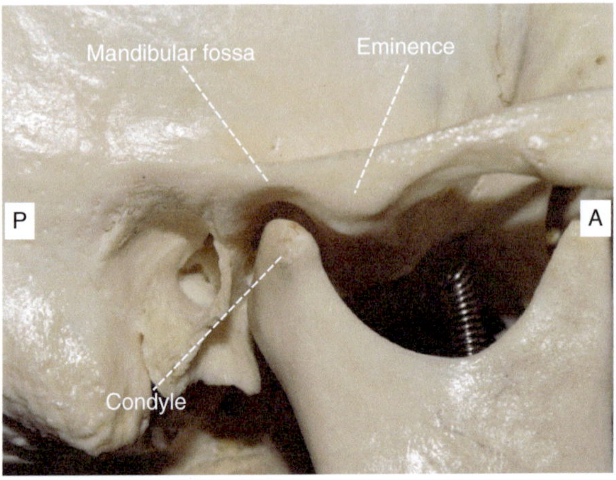

Fig. 18.1 Temporomandibular joint (right)

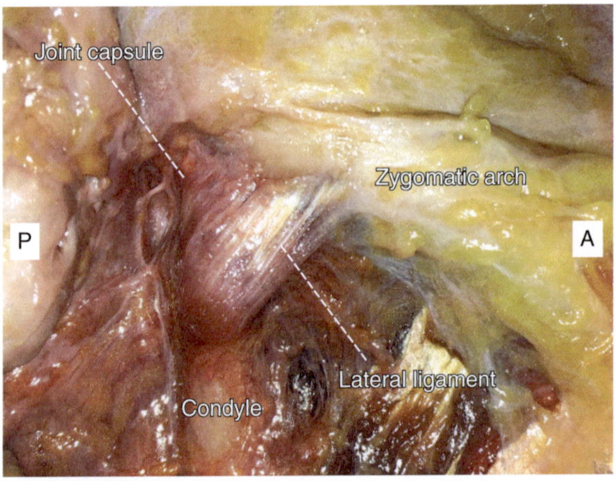

Fig. 18.2 Fibrous capsule and lateral ligament (right)

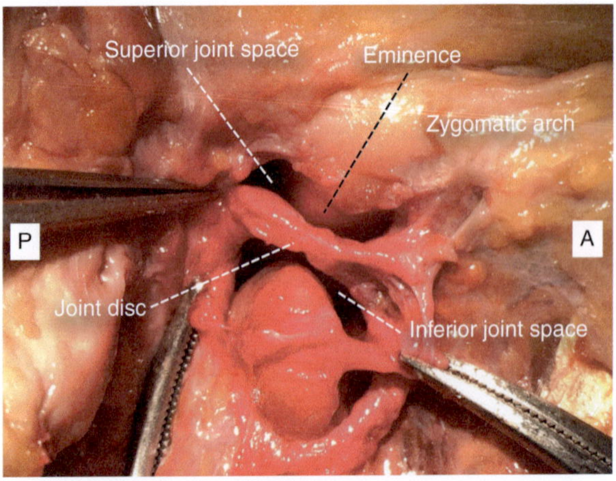

Fig. 18.3 Joint space. Note the joint disc separating the space into two parts (right)

vascularized, forming capillary networks over the synovial membranes. From a clinical perspective, understanding the anatomy and variations of the TMJ is significant for TMJ surgery, diagnosis of temporomandibular joint disorder, and related diseases and treatments.

18.2 Normal Anatomy and Variations of the Temporomandibular Joint

Variations of the TMJ involve differences in size, shape, or a compilation of the different components. The anomalies are deviations from what is normally found and is asymptomatic. Discrepancies in joint part sizes are usually proportional to general skeletal size; however, there can be structural anomalies that are disproportionate to the skeleton. Some examples include small condyles in large fossae, tight effective attachment of the disc to the condyle with loose attachment to the temporal bone or vice versa, strong joint ligaments, and steeply inclined articular eminences. A functional false joint anterior to the glenoid fossa, large and irregular articulating surfaces, false condyles with medial thickening, a shallower glenoid fossa, and a shorter postglenoid process have been found in the region of the TMJ. In rare cases the entire TMJ is absent. This usually presents asymptomatically since the joint is non-load-bearing.

18.2.1 Muscles Related to the TMJ

The muscles of mastication attached to the TMJ are responsible for mastication. These muscles can be hypertrophied or even absent as a result of congenital defects. A study of the insertion of the pterygoid muscle into the disc-capsule complex of the TMJ revealed two-headed and three-headed variants (Kilic et al. 2010). Another investigations into the attachment of superior head of the lateral pterygoid muscle found three variant types of attachment: the muscle inserting into the condyle and the disc-capsule complex (type I: 36.7-55.5%), the muscle inserting only into the condyle (type II: 26.5-27.8%), and the muscle inserting only into the disc-capsule complex (type III: 5-16.7%) (Naidoo 1996; Kilic et al. 2010; Antonopoulou et al. 2012) (Fig. 18.4).

18.2.2 Condyle

There are individual variations in the shape, form, position, presence, and size of the mandibular condyle. Reports indicate that the prevalence of radiographic variants in condylar morphology is 81.3% (Mathew et al. 2011). The condyle is commonly classified into five groups: normal, flattening, sclerosis, osteophytes, and erosion (Fig. 18.5). It has been documented as 15–20 mm in length and 8–10 mm in width, and it can protrude 15–20 mm medially. In rare cases, the mandibular condyle is superiorly convex.

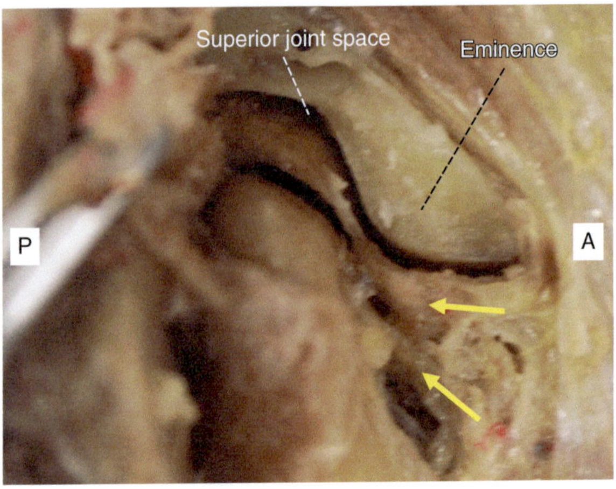

Fig. 18.4 Insertion of the lateral pterygoid muscle (arrows) (right)

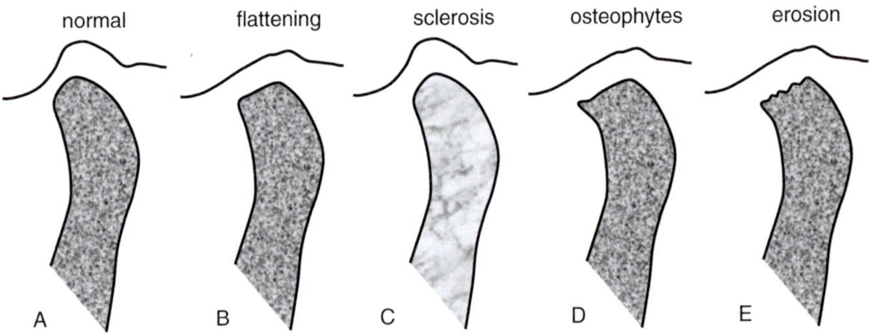

Fig. 18.5 Condyle classification

The neck of the condyle can vary in length and width, with a central ridge on the anterior surface of the neck. There is sometimes a fovea on the anterior surface of the neck of the condyle, where the lateral pterygoid muscle can insert. The condyle can also have a backward or forward inclination with variations in the length of the neck. Some condyles have a smooth continuation of the anterior neck into the convex articular surface, while others have a narrow depression where the head of the condyle meets the neck, coexisting with a sharp anterior protrusion of the head (Zarb and Carlsson 1979). The condylar head can grow vertically, horizontally, or rotationally, changing the position of the joint. The mandibular condyle can also be higher than the coronoid process, and its angle in relation to the ramus can vary.

18.2.2.1 Condylar Posterior Surface Concavity

Variations in shape are common radiographic findings in the posterior surface of the condyle. The bony surface of the condyle can be flattened or contain a bony protuberance that can cause an anatomical perforation of the disc. Although normally arched and concave, the condyle can be convex or have a smooth depression in the superior articular, which can be mistaken for a bifid condyle. The concavity can also vary in extent.

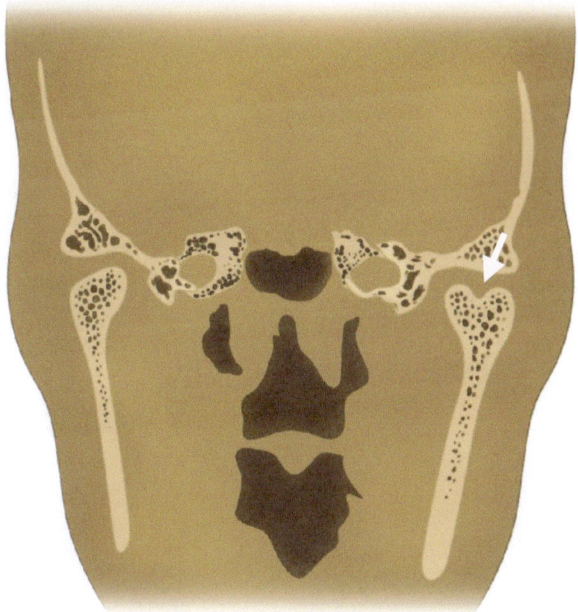

Fig. 18.6 Bifid condyle

18.2.2.2 Multilobular Condyle

Bifid condyles (double condyle) result from replication of the head of the condyle, usually in the anteroposterior aspect (Fig. 18.6). These are rare and most present asymptomatically. Bifid condyles can arise from separation of the lateral and medial aspects of the developing condyle by the vascular canals and their connective tissue septa that form in it (Moffett 1966). The splitting of the condyle can range from a shallow groove to two distinct condyles (Stefanou et al. 1998). A medial and a lateral head divided by an anteroposterior groove or an anterior and posterior head can be present. A multilobular condyle is usually unilateral but sometimes bilateral. It is more likely to be diagnosed incidentally by radiological rather than clinical evidence. A study of 1882 cadaveric skulls reported the incidence of bifid mandibular condyles as 0.48% (Szentpetery et al. 1990). Even rarer than the bifid condyle, cases of trifid (three-lobed) and tetrafid (four-lobed) condyles have been reported (Sala-Perez et al. 2010; Sahman et al. 2011). One case report demonstrated congenital Frey's syndrome associated with a bilateral trifid mandibular condyle. The patient had no history of trauma or surgery (Motta-Junior et al. 2013). Complete absence of the left mandibular condyle has been reported with no previous history of trauma or disease (Canger and Celenk 2012).

18.2.3 Mandibular Fossa

The mandibular condyle is located within the mandibular fossa (also called the glenoid fossa) of the temporal bone. The thinnest part of the roof of the fossa is reported

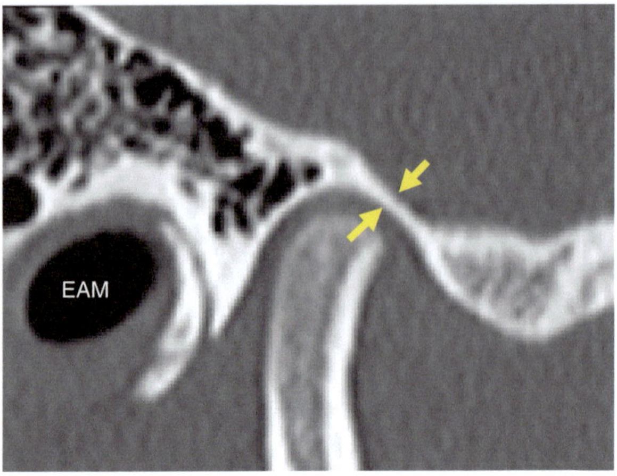

Fig. 18.7 The thinnest part of the roof of the fossa in sagittal section on CT (right). *EAM* external acoustic meatus

to be 1.0–1.2 mm (Al-Koshab e al. 2015) (Fig. 18.7). Anatomically, this thin structure could entail some risk for the pumping technique in treating TMJ disorder with closed lock and autologous blood injection, given the habitual dislocation of the TMJ. The tip of the needle could perforate the roof and go into the middle cranial fossa. The presence of fibrocartilage rather than hyaline cartilage is important because it has the ability to repair and remodel. The thinnest portion of the mandibular fossa may be lined with synovial membranes.

18.2.4 Articular Eminence

The articular eminence and articular surface are approximately 25° to the occlusal plane but the steepness varies. The eminence is somewhat saddle-shaped, strongly convex from the side, and concave when viewed from the front or back. The degree of this convexity and concavity is highly variable and is considered to be related to dislocation of the TMJ. However, both steep and gentle eminences have been reported as a cause of dislocation. Many recent studies show that eminectomy, which makes the eminence flat, is a better choice for treating habitual dislocation of the TMJ than other invasive surgeries. Fine bony ridges that are not secondary to a disease process or trauma often accentuate the medial and lateral borders of the articular eminence. Disc displacement and disc position can result directly from the steepness of the articular eminence. The eminence can be less prominent or absent from the TMJ; its loss can result in a widened disc space.

Occasionally, the mastoid cells of the temporal bone extend to the zygomatic arch (Carter et al. 1999) (Fig. 18.8). In terms of eminectomy for treating habitual dislocation of the TMJ or other invasive surgery for the posterior part of the zygomatic arch, the surgery could invade the extended mastoid cells, and mastoiditis could ensue.

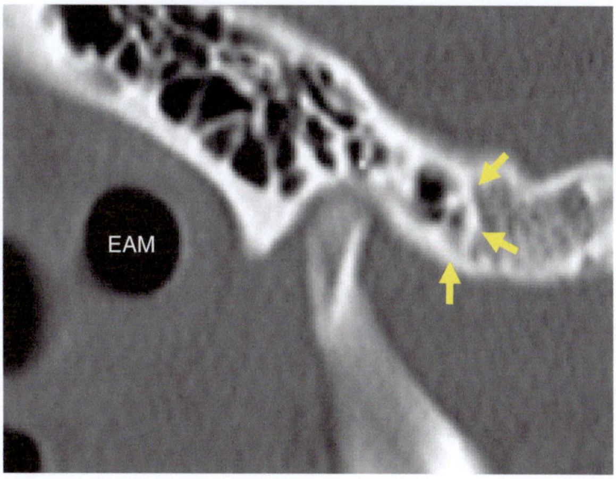

Fig. 18.8 Extended mastoid cells to the zygomatic arch (arrows) in sagittal section on CT. *EAM* external acounstic meatus

18.2.5 Coronoid Process

18.2.5.1 Coronoid Process Hyperplasia

Hyperplasia of the coronoid process is infrequent and can present with difficulty or limitation in opening the mouth. It is more prevalent in females. Bilateral cases are five times more common than unilateral cases (Neville et al. 2008). There can also be nodular growth of the tip of the coronoid process.

18.2.5.2 Elongation of the Coronoid Process

A geriatric male with right and left coronoid process lengths of 2.4 cm and 2.6 cm, respectively, has been reported (Chauhan and Gupta 2011).

18.2.6 Ramus of the Mandible

18.2.6.1 Absence of the Ramus of the Mandible

Agenesis can entail partial to total absence of the TMJ due to absence of the condyle or rami of the mandible. Congenital bilateral absence of the mandibular rami has also been noted (Kazanjian 1956). Persons with unilateral agenesis or any other form of unilateral mandibular growth arrest will demonstrate a very characteristic appearance: a long face on one side and a short face on the other. The facial asymmetry includes mandibular deviation to the affected side when the mouth is opened.

18.2.7 Maxilla

18.2.7.1 Union of the Coronoid Process and Maxilla

There have been reports of a bilateral union between the coronoid processes and the maxilla (Super and Cotton 1982).

18.2.7.2 Synostosis of Mandible and Maxilla

Jaw dysfunction can result from hypertrophy of the coronoid process. If this hypertrophy occurs bilaterally, the interference is spontaneous. A patient has been reported with pseudoankylosis due to fusion between the lingula of the mandible and the left pterygoid process (Super and Cotton 1986). Other cases presented with limited ability to open the mouth, caused by fusion of the maxilla and mandible (Shams et al. 2006) and the maxilla and mandible and zygomatic bone (Fallahi et al. 2010).

18.2.8 Disc

The articular disc, also referred to as the meniscus, is composed of avascular dense fibrous connective tissue with some chondrification in areas of maximum loading, positioned between the mandibular condyle and mandibular fossa of the temporal bone. In sagittal section, this articular disc has three distinct parts: anterior band, thinner intermediate zone, and posterior band. The intermediate zone lacks vascularity and acts as the force-accepting segment. Because of this structure, the disc is biconcave and appears as a "bow tie" in sagittal section on MRI (Fig. 18.9).

The normal thickness falls within the ranges 2–3 mm posteriorly and 1–2 mm in the middle. The shape of the articular surfaces and the condyle usually determines the size and shape of the disc: The steeper the articular eminence, the greater the posterior thickness of the disc. Normally, the disc intervenes between the head of the condyle and the mandibular fossa and is thinnest at the posterior slope of the articular tubercle.

18.2.8.1 Disc Displacement (Internal Derangement)

A centric condyle position is considered the correct interposition of the disc; however, the disc can also have a significant anterior or posterior position. The condyle can also be anterior or posterior relative to a normal disc position (Isberg 2001). If the anterior

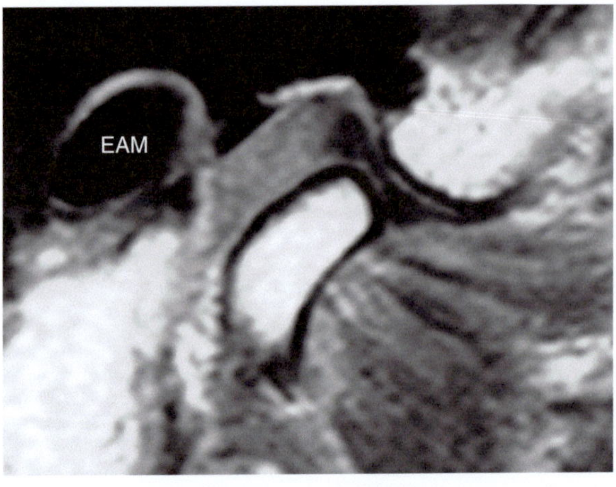

Fig. 18.9 The disc appears as a "bow tie" in sagittal section on MRI (right). *EAM* external acoustic meatus

portion of the condyle and the inferior part of the disc are >2 mm unaligned, the disc is considered displaced, which means that the anterior prominence of the condyle articulates against the posterior band of the disc or the side of the disc. The disc can be partially or completely displaced, either all or part of it not being found in the normal position. A disc can be displaced in any direction (anterior, posterior, lateral, medial, anterolateral, anteromedial, posterolateral, or posteromedial). Medial disc displacement can present with the auriculotemporal nerve trunk coursing between the bony joint components in the articular fossa, instead of close to the condyle at the level of the temporal bone. The disc can also be detached from the superior retrodiscal tissue or be trapped anterior to the condyle with no superior joint space.

Mild dislocation, or subluxation, can result from variations in disc tautness or looseness. Generalized joint laxity has been classified and cited as a normal variant and is significantly more prevalent in patients with internal derangements (Magnusson et al. 1994). Joint laxity can result from lengthening of the anterior recess and a drag on the inferior surface of the disc during the rotation phase of early mouth opening or from changes in ligament structure (Meunissier et al. 1993).

A study of asymptomatic volunteers reported a disc displacement prevalence of 33% (Katzberg et al. 1996). A study of disc position in asymptomatic children with average age 11 years reported a rate of 6%. The prevalence increased to 34% in a population with a median age of 16–19 years. These findings suggest that children begin to develop changes in disc position during their early teenage years (Tominaga et al. 2007).

18.2.8.2 Joint Space

The width of the joint space (the distance between the condyle and its fossa) can vary. If there is hyperplasia of the condyles, the joint space tends to be narrower, whereas hypoplasia of the condyles tends to present with wider joint spaces. The joint cavity is divided into two (superior and inferior joint spaces) by an articular disc (Fig. 18.3). The volumes of the superior and inferior joint space with mouth opening are in the ranges 1.2–3.2 cc and 0.7–1.5 cc, respectively.

18.2.8.3 Perforation of the Disc

Mechanical breakdown of the articular disc can lead to perforation (instead of unusual disc position) (Fig. 18.10), which in turn causes chronic disc displacement and thickening (Wongwatana et al. 1994). The disc can be perforated at the posterior slope of the articular tubercle with variations at the articular eminence. Such perforation can result in noises from the TMJ, such as crepitation.

A study of perforations of the TMJs from 106 cadavers reported disc perforation in 28% with an age range of 53–90 years. The perforations were located centrally, laterally, and posteriorly. Central perforation was found in 48.8% of all perforations, followed by lateral perforation (41.9%) and posterior perforation (9.3%). Central perforations were large, almost causing complete destruction of the disc, and there was severe condylar remodeling under the perforation. In contrast, lateral perforations ranged from small to large, and the associated condylar changes varied with the size of the defect. Regardless of the amount of destruction, the discs remained well attached to the periphery of the condyle (Barton and Ellenbecker 1987).

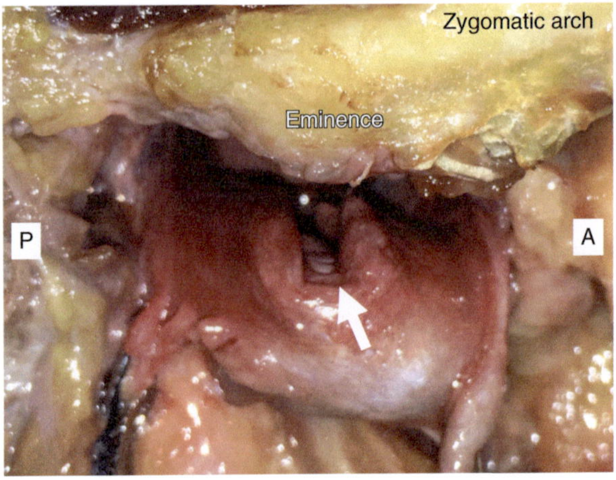

Fig. 18.10 Perforation of the disc

18.2.8.4 Disc–Condyle Adhesions

Fibrous adhesions within the TMJ that are not caused by trauma or congenital deformities are thought to occur mainly in the superior compartment of the TMJ, specifically between the disc and the condyle. Bilateral adhesions can cause deepening of pre-angular notches (Bell 1990). Intra-articular disc adhesions often accompany thickening and flattening of the condylar head and narrowing of the joint space.

18.2.9 Joint Capsule

The TMJ is completely surrounded by a joint capsule from the articular margins of the mandibular fossa to the neck of the mandible. The joint capsule encloses the articular disc at the inner aspect of the capsule. Short capsular fibers that run from the disc to the neck of the condyle support the TMJ. This arrangement allows the disc to be loose and the portion of the capsule below it to be tight. Sesamoid bones can occur in the TMJ (Lang 1995). Two synovial membranes are associated with the TMJ, one located superior and the other inferior to the disc. They both line the non-articular surfaces and fuse with the periphery of the disc.

The TMJ capsule can be stretched or its size can vary. Variations in the capsule often present with chronic dislocations of the disc.

18.2.10 Related Ligaments

The TMJ consists of three ligaments: lateral, sphenomandibular, and stylomandibular. Two of these are accessory and guide the movement of the mandible against the base of the skull. All of the ligaments are believed to be under the influence of muscular activity.

18.2.10.1 Lateral Ligament

The lateral ligament reinforces the lateral aspect of the joint capsule, which extends posteroinferiorly at an angle of approximately 45° to the horizontal (see Fig. 18.2). It attaches to the lateral surface and posterior border of the neck of the condyle deep to the parotid gland. It may limit lateral displacement of the TMJ.

18.2.10.2 Sphenomandibular Ligament

This is found on the medial side of the joint and courses from the spine of the sphenoid to the lingula of the mandible. It is thin and functions as the primary passive support of the mandible. The attachment of the sphenomandibular ligament could be duplicated (Simonds et al 2017).

18.2.10.3 Stylomandibular Ligament

This is a thickening of the deep cervical fascia that separates the parotid and submandibular glands. It runs from the styloid process of the temporal bone to the posterior border of the ramus of the mandible.

18.2.10.4 Other Ligaments

There are cases of an extra ligament—the mandibular-malleolar or anterior malleolar ligament—connecting the neck and anterior process of the malleus to the superomedial part of the posterior capsule, interarticular disc, and sphenomandibular ligament (Pinto 1962). Variations in the lateral ligament of the TMJ have also been described. A macroscopic investigation revealed that 70% of 20 TMJ specimens exhibited an obvious ligament and the remaining 30% exhibited an indistinct structure. Of the obvious ligaments, 35.7% formed a cord-like structure with sharp borders (Nell et al. 1994).

18.2.11 Blood Vessels

18.2.11.1 Arterial Supply

The maxillary artery is the largest terminal branch of the external carotid artery. It arises posterior to the neck of the mandible and runs deep to the mandibular condyle and lateral to the stylomandibular ligament. The TMJ obtains its arterial blood supply from the middle meningeal and anterior tympanic arteries. The superficial temporal artery is another terminal branches of the external carotid, which emerges on to the face between the TMJ and the auricle. It runs with the auriculotemporal nerve. The ascending pharyngeal artery emerges as the first or second branch and the smallest branch of the external carotid artery, which ascends deep to the internal carotid artery. The masseteric artery accompanies the masseteric nerve and supplies branches to the TMJ. The deep auricular artery also contributes a small branch to the TMJ.

The ascending pharyngeal artery can branch from the common carotid instead of the external carotid (Standring 2009). There are variations of the superficial temporal artery involving the relative sizes of branches and their absence (Standring 2009). The superficial temporal and middle meningeal arteries can also be superimposed.

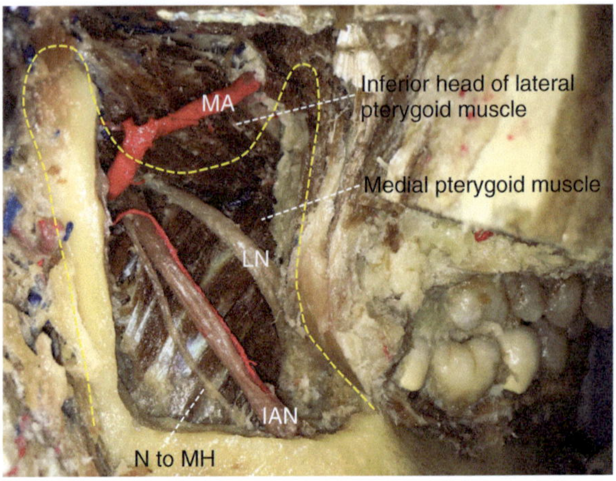

Fig. 18.11 Course of the maxillary artery (most of the ramus of the mandible has been removed). Note the maxillary artery in this picture runs lateral to the lateral pterygoid muscle. *IAN* inferior alveolar nerve, *LN* lingual nerve, *MA* maxillary artery, *N to MH* nerve to the mylohyoid

There are many known variations in the course of the maxillary artery (Fig. 18.11). These include passage of the artery medial to the lateral pterygoid and to the lingual and inferior alveolar nerves, passage between the lingual and inferior alveolar nerves, passage through a loop formed by the inferior alveolar nerve, and branching of the middle meningeal artery distal to the inferior alveolar artery. The maxillary artery can pierce different aspects of the lateral pterygoid and can also present superficial or deep to the lower head of the lateral pterygoid. The main stem of the maxillary artery can run superficial to both heads of the lateral pterygoid muscle or pass deep to its inferior head and emerge through the cleft between the two heads. The relationship between the lateral pterygoid muscle and the maxillary artery differs among populations. The incidence of an internal course (medial to the lateral pterygoid) has been reported as 3.6–10.1% in Japan, 44.7% in the United Kingdom, 45.6% among Caucasians in the United States, and 91.5% in Australia (Iwanaga et al. 2017). The maxillary artery has also been found to arise from a common stem with the facial artery, which allows it to be parallel with the ramus of the mandible.

The middle meningeal artery can arise either directly from the first part of the maxillary artery or from a common trunk with the inferior alveolar artery. The first alternative is more likely when the maxillary artery lies superficial to the lateral pterygoid. Variations in the origin of the middle meningeal artery in relation to the inferior alveolar artery have also been noted. Penetrating vessels that supply the lateral pterygoid can also supply the mandibular condyle.

Vascular malformation is rare in the temporomandibular joint as two-thirds of malformations of the jaw occur in the mandible, particularly the ramus or mandibular body. Most of these malformations are asymptomatic with clinical, microscopic, and radiographic variations that create a major challenge for the accurate diagnosis of patients (Abramowicz et al. 2007).

18.2.11.2 Venous Drainage

The maxillary vein, which drains the pterygoid plexus, unites with the superficial temporal vein to form the anterior division of the retromandibular vein, a deep vessel of the face that runs posterior to the ramus of the mandible within the parotid gland and superficial to the external carotid artery to drain the TMJ. The posterior division of the retromandibular vein and the posterior auricular veins form the external jugular vein, which drains the TMJ.

The retromandibular vein is often not connected to the posterior auricular vein to form the external jugular, enlarging the anterior jugular vein. The veins drain the anterior aspect of the joint and its associated tissues into the plexus that surrounds the lateral pterygoid.

18.2.11.3 Lymphatic Drainage

The TMJ is drained by the superficial parotid lymph nodes and the lateral cervical lymph nodes.

18.2.12 Nerve Supply

The mandibular nerve arises from the trigeminal nerve (CN V). It gives two branches, the auriculotemporal and masseteric nerves, to the TMJ. The auriculotemporal nerve usually encircles the middle meningeal artery and courses medial to the neck of the mandible and posterior to the TMJ. It sends articular branches to the TMJ. The masseteric nerve also provides articular branches to the TMJ.

The nerves to the temporal muscles (there can be one or three nerves, but usually there are two, the anterior and posterior deep temporal) either arise from the upper head of the lateral pterygoid muscle and the anterior bone forming the infratemporal fossa or emerge between the two heads of the lateral pterygoid muscle. Three spurious jugular foramina have been reported in the glenoid fossa of the temporomandibular joint and three in the zygomatic process (Angel 1948). The deep temporal nerves are usually accompanied by corresponding deep temporal arteries, but these again vary in number. The anterior deep temporal nerve can either follow the masseteric nerve above the upper border of the superior head of the lateral pterygoid muscle and the bone or emerge between the two heads of the lateral pterygoid. The masseteric and deep posterior temporal nerves lie 2–5 mm posterior to the articulating eminence.

18.3 Congenital Anomalies

Congenital anomalies of the TMJ include hyperplasia, hypoplasia, aplasia, and dysplasias. Familial traits of bilateral retrognathism or prognathism are not uncommon, whereas bilateral temporomandibular maldevelopments such as Treacher Collins syndrome are very rare.

18.3.1 Hyperplasia

Some hemihyperplasia (hemihypertrophy) syndromes include Beckwith–Wiedemann syndrome, Klippel–Trenaunay–Weber syndrome, McCune–Albright syndrome, Langer–Giedion syndrome, Maffucci syndrome, and Ollier disease. These syndromes and disease are usually unilateral and result in movement disorders of the TMJ.

18.3.2 Hypoplasia

Hypoplasia syndromes include mandibular synostosis, oculo-auriculo-vertebral syndrome (Goldenhar syndrome), and hemifacial microsomia. Hemifacial microsomia, first and second branchial arch syndromes, or lateral facial dysplasias are commonly characterized as affecting the condyle and ramus of the mandible as well as the temporal bone. Hemifacial microsomia, or first arch syndrome, is characterized by mandibular growth with insufficient soft tissue, hypoplasia of the first and second brachial arch muscles, and facial nerve palsy. The ramus of the mandible and TMJ classify this skeletal defect. It can present clinically with small sizes, hypoplastic anterior and medially displaced TMJ, or complete absence of the joint.

Treacher Collins syndrome, or mandibulofacial dysostosis, entails bilateral facial microsomia and presents with hypoplastic TMJs with short mandibular rami. It includes craniofacial anomalies such as incomplete development of the cranial bones and the first and second brachial arches of the mandible. In Treacher Collins syndrome, the mandible is smaller and the condyle will present smaller with posterior and inferior positioning in relation to the external auditory canal.

Hallermann–Streiff syndrome, also known as oculomandibulodyscephaly, presents with a small mandible, especially the condyles, and an anterior positioning of the condyle with a poorly formed fossa. It can be differentiated from Treacher Collins syndrome by the articulation of the condyles. In both syndromes, the condyles are hypoplastic; however, Treacher Collins syndrome presents with hypoplastic condyles that have a normal anteroposterior relationship to the fossa, while Hallermann–Streiff syndrome presents with hypoplastic condyles articulating with the articular eminence.

The Robin anomaly and Pierre–Robin syndrome feature hypoplasia of the mandible and maxilla.

18.3.3 Aplasia and Agenesis

Aplasia and agenesis are most commonly unilateral, but asymmetry is common in bilateral cases. Goldenhar syndrome exhibits partial or total absence of the TMJ. Nager–Reynier syndrome, or acrofacial dysostosis, is an abnormality of the TMJ that closely resembles those found in facial microsomia, but maldevelopment is limited to the TMJ (which can be hypoplastic or absent) and the ramus of the mandible.

18.3.4 Dysplasias

Fibrous dysplasia occurs as a slow benign swelling of the mandible or maxilla, where the fibrous connective tissue has a characteristic whorled pattern and contains trabeculae of immature nonlamellar bone. Fibrous dysplasia occurs in children and young adults and usually becomes inactive when they reach skeletal maturity.

References

Abramowicz S, Marshall CJ, Dolwick MF et al (2007) Vascular malformation of the temporomandibular joint: report of a case and review of the literature. Oral Surg Oral Med Oral Path Oral Radiol 103:203–206. https://doi.org/10.1016/j.tripleo.2006.05.002

Al-Koshab M, Nambiar P, John J (2015) Assessment of condyle and glenoid fossa morphology using CBCT in south-east Asians. PLoS One 10(3):e0121682. https://doi.org/10.1371/journal.pone.0121682

Angel JL (1948) Factors in temporomandibular joint form. Am J Anat 83:223–246

Antonopoulou M, Iatrou I, Paraschos A et al (2012) Variations of the attachment of the superior head of human lateral pterygoid muscle. J Craniomaxillofac Surg 41:1–7. https://doi.org/10.1016/j.jcms.2012.11.021

Barton JM, Ellenbecker MA (1987) Anatomical variations in the articular disc of the human temporomandibular joint. Trans Nebraska Acad Sci XV:1–4

Bell WE (1990) Temporomandibular disorders: classification, diagnosis, management, 3rd edn. Year Book Medical Publishers, Chicago

Canger EM, Celenk P (2012) Aplasia of the mandibular condyle associated with some orthopaedic abnormalities: a case report. Dentomaxillofac Radiol 31:259–263. https://doi.org/10.1259/dmfr/93380292

Carter LC, Haller AD, Calamel AD et al (1999) Zygomatic air cell defect (ZACD). Prevalence and characteristics in a dental clinic outpatient population. Dentomaxillofac Radiol 28:116–122

Chauhan P, Gupta S (2011) Bilateral elongated coronoid processes of mandible. Int J Anat Var 4:25–27

Fallahi HR, Naeini M, Mahmoudi M et al (2010) Congenital zygomatico-maxillo-mandibular fusion: a brief case report and review of literature. Int J Oral Maxillofac Surg 39:930–933

Isberg A (2001) Temporomandibular joint dysfunction: a practitioner's guide. Informa Healthcare, London

Iwanaga J, Kikuta S, Nakamura M et al (2017) Intraoral vertico-sagittal ramus osteotomy: modification of the L-shaped osteotomy. Int J Oral Maxillofac Surg. https://doi.org/10.1016/j.ijom.2017.06.003

Katzberg RW, Westesson PL, Tallents RH et al (1996) Anatomic disorders of the temporomandibular joint disc in asymptomatic subjects. J Oral Maxillofac Surg 54:147–153

Kazanjian VH (1956) Bilateral absence of the ascending rami of the mandible. Br J Plast Surg 9:77–82

Kilic G, Dergin G, Yazar F et al (2010) Insertions of the lateral pterygoid muscle to the disc-capsule complex of the temporomandibular joint and condyle. Turk J Med Sci 40:435–441. https://doi.org/10.3906/sag-0808-29

Lang J (1995) Clinical anatomy of the masticatory apparatus and Peripharyngeal spaces. Thieme, New York

Magnusson T, Carlsson GE, Egermark I (1994) Changes in clinical signs of craniomandibular disorders from the age of 15 to 25 years. J Orofac Pain 8:207–215

Mathew AL, Sholapurkar AA, Pai KM (2011) Condylar changes and its association with age, TMD, and dentition status: a cross-sectional study. Int J Dent 2011:1–7. https://doi.org/10.1155/2011/413639

Meunissier M, Meunier A, Carpentier P et al (1993) Disc movements over the condylar head: radiographical study on autopsy materials. J Orofac Rehabil 20:501–515
Moffett B (1966) The morphogenesis of the temporomandibular joint. Am J Orthod 52:401–415
Motta-Junior J, Aita TG, Pereira-Stabile CL et al (2013) Congenital Frey's syndrome associated with nontraumatic bilateral trifid mandibular condyle. Int J Oral Maxillofac Surg 42:237–239. https://doi.org/10.1016/j.ijom.2012.06.016
Naidoo LCD (1996) Lateral pterygoid muscle and its relationship to the meniscus of the temporomandibular joint. Oral Surg Oral Med Oral Pathol Oral Radiol Endod 82:4–9
Nell A, Niebaurer G, Sperr W et al (1994) Special variations of the lateral ligament of the human TMJ. Clin Anat 7:267–270. https://doi.org/10.1002/ca.980070507
Neville BW, Damm DD, Allen C, Bouquot J (2008) Oral and maxillofacial pathology, 3rd edn. Saunders, Philadelphia
Pinto O (1962) A new structure related to the temporomandibular joint and the middle ear. J Prosthet Dent 12:95–103. https://doi.org/10.1016/0022-3913(62)90014-8
Sahman H, Etoz OA, Sekerci AE et al (2011) Tetrafid mandibular condyle: a unique case report and review of the literature. Dentomaxillofac Radiol 40:524–530. https://doi.org/10.1259/dmfr/62082661
Sala-Perez S, Vazquez-Delgado E, Rodriguez-Baeza A et al (2010) Bifid mandibular condyle: a disorder in its own right. J Am Dent Assoc 131:1076–1085
Shams MG, Motamedi MHK, Abad HLD (2006) Congenital fusion of the maxilla and mandible: brief case report. Oral Surg Oral Med Oral Path Oral Radiol 102:e1–e3. https://doi.org/10.1016/j.tripleo.2005.10.051
Simonds E, Iwanaga J, Oskouian RJ et al (2017) Duplication of the Sphenomandibular Ligament. Cureus 9:e1783. https://doi.org/10.7759/cureus.1783
Standring S (2009) Gray's anatomy, 40th edn. Churchill Livingstone, London
Stefanou EP, Fanourakis IG, Vlastos K et al (1998) Bilateral bifid mandibular condyles: report of four cases. Dentomaxillofac Radiol 27:186–188. https://doi.org/10.1038/sj/dmfr/4600343
Super S, Cotton JS Jr (1982) Bilateral pseudoankylosis of the TMJ due to synostoses between the mandible and maxilla. J Oral Maxillofac Surg 40:590–592
Super S, Cotton JS Jr (1986) A case of pseudoankylosis between the pterygoid plate and mandible. J Oral Maxillofac Surg 44:467–468
Szentpetery A, Gabor K, Marcsik A (1990) The problem of the bifid mandibular condyle. J Oral Maxillofac Surg 48:1254–1257
Tominaga K, Konoo T, Morimoto Y et al (2007) Changes in temporomandibular disc position during growth in young Japanese. Dentomaxillofac Rad 36:397–401. https://doi.org/10.1259/dmfr/40410443
Tubbs RS, Shoja MM, Loukas M (2008) Letter to the editor. Clin Anat 21:609
Wongwatana S, Kronman JH, Clark RE et al (1994) Anatomic basis of disc displacement in temporomandibular joint dysfunction. Am J Orthod Dentofacial Orthop 105:257–264
Zarb GA, Carlsson GE (eds) (1979) Temporomandibular joint: function and dysfunction. Mosby, Missouri

Part VIII

Teeth

Variations in the Number of Teeth

19

Tsuyoshi Tanaka

19.1 Introduction

There are two main congenital anomalies of human tooth number: hyperdontia and congenitally missing teeth. Hyperdontia is the condition of having more teeth than the normal set (28, excluding third molars); congenitally missing teeth means fewer teeth than normal. Both conditions can appear as isolated traits or associated with syndromes such as ectodermal dysplasia. Tooth number variations can have negative aesthetic and functional effects such as malocclusion, periodontal injury, impaired alveolar bone growth, reduced chewing efficiency, and speaking problems. Patients who have tooth number irregularities need more dental attention, so these conditions are interesting not only for dentists but also for public health departments and health insurance companies.

19.2 Hyperdontia (Supernumerary Teeth)

19.2.1 Definition

Hyperdontia is the condition of having more teeth than normal in both permanent and primary dentitions. It occurs in one or both jaws and can be either single or multiple, unilateral or bilateral, and erupted or impacted. There are two categories of hyperdontia, supplemental teeth and supernumerary teeth. A supplemental tooth is normally shaped with no anatomical or functional difference from the norm. A supernumerary tooth is different from a normal tooth in shape and size. It does not resemble any tooth with which it is associated. Supernumerary teeth are further distinguished as conical, tuberculate, or infundibular in shape. Conical supernumerary

T. Tanaka (✉)
University of Florida, Gainesville, FL, USA

J. Iwanaga, R. S. Tubbs (eds.), *Anatomical Variations in Clinical Dentistry*,
https://doi.org/10.1007/978-3-319-97961-8_19

teeth are most commonly found in the permanent dentition (Rajab and Hamdan 2002). The tuberculate form is barrel-shaped and displays more than one cusp and an incomplete or entirely missing root. Infundibular teeth can always be found in the upper incisor area, and they have a crown with funnel in the occlusal surface. Their root is single and tapered.

19.2.2 Prevalence

The incidence of hyperdontia in permanent dentition is reported as 1.5–3.5% and in primary dentition as 0.3–0.8% (Garvey et al. 1999; Mahabob et al. 2012). We often overlook hyperdontia in the primary dentition because the teeth are similar to normal shape and erupt in normal alignment. They can be mistaken for gemination or fusion anomalies. Males are affected twice as frequently as females (Primosch 1981). Hyperdontia occurs 8.2 times more frequently in the maxilla than the mandible in the permanent dentition. Several terms have been used to describe supernumerary teeth according to their location. Mesiodens is the maxillary anterior incisor region and this is most commonly found. Paramolar is lingual or buccal to a molar tooth, and distomolar/distodens is distal to the third molar. According to Gomes et al. (2008), the most common supernumerary teeth in Brazilian children and adolescents are maxillary central incisors (Figs. 19.1 and 19.2).

Listed in order of frequency are the following:

1. Maxillary central incisor (48.3%).
2. Maxillary midline (mesiodens) (28.5%).
3. Maxillary lateral incisor (10.0%).
4. Mandibular premolar (6.5%) (Fig. 19.3).
5. Maxillary canine (2.6%).

A detailed description of supernumerary tooth location is shown in Fig. 19.4.

Most commonly there is only one supernumerary tooth in a dentition; less frequently there are two supernumeraries, and more supernumerary teeth are rare. The prevalence of multiple supernumerary teeth is 0.06% (Acikgoz et al. 2006).

19.2.3 Etiology

Although there are many hypotheses, the etiology of hyperdontia is still unknown. This problem seems to be caused by genetic or environmental factors. Supernumerary teeth could result from the development of excess dental lamina and additional tooth germ formation or could develop from separation of the tooth bud. Taylor (1972) stated that the tooth bud splits into two equal or different-sized parts, resulting in two teeth of equal size or one normal and one dysmorphic tooth. Another suggested cause is hyperactivity of the dental lamina, where the epithelial cells that form supernumerary teeth remain for long periods (Rajab and Hamdan 2002). Heredity could also affect the prominence of this condition. In many cases, supernumerary teeth such as mesiodens recur within the same family (Bailleul-Forestier

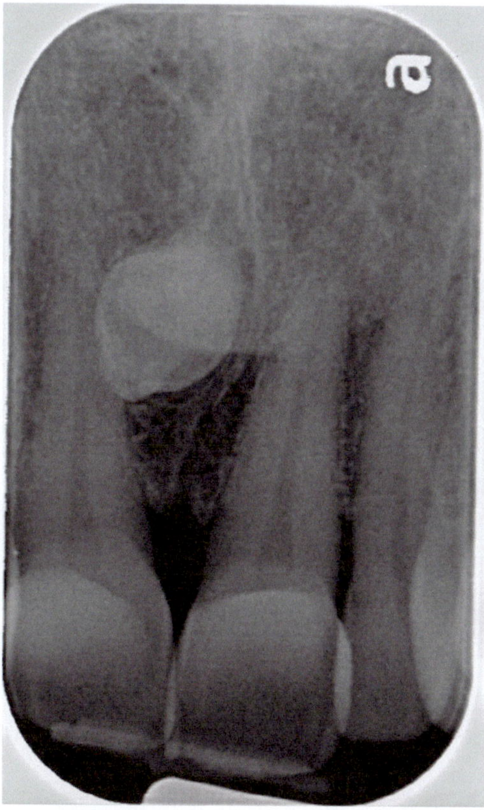

Fig. 19.1 Mesiodens (Courtesy of Dr. Nicole Martino)

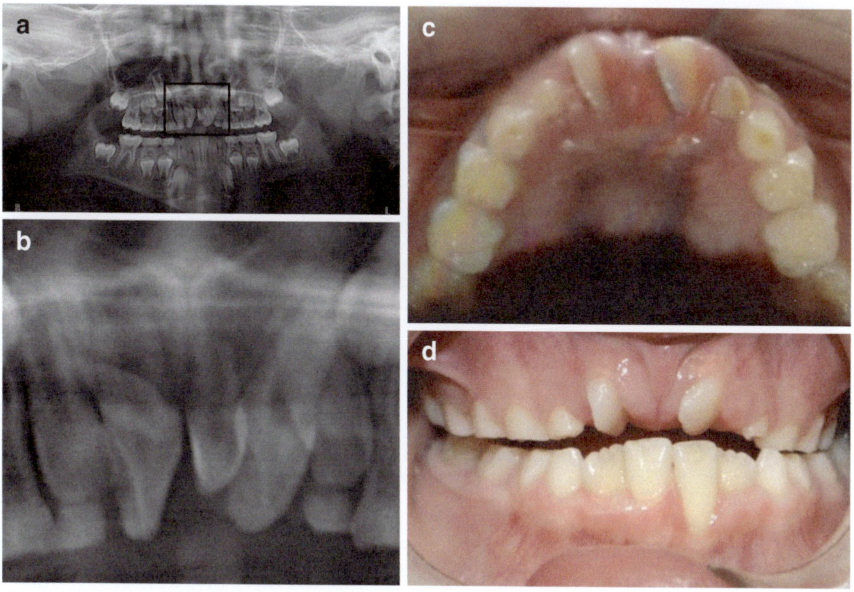

Fig. 19.2 Mesiodens (Courtesy of Dr. Christopher L. Connell)

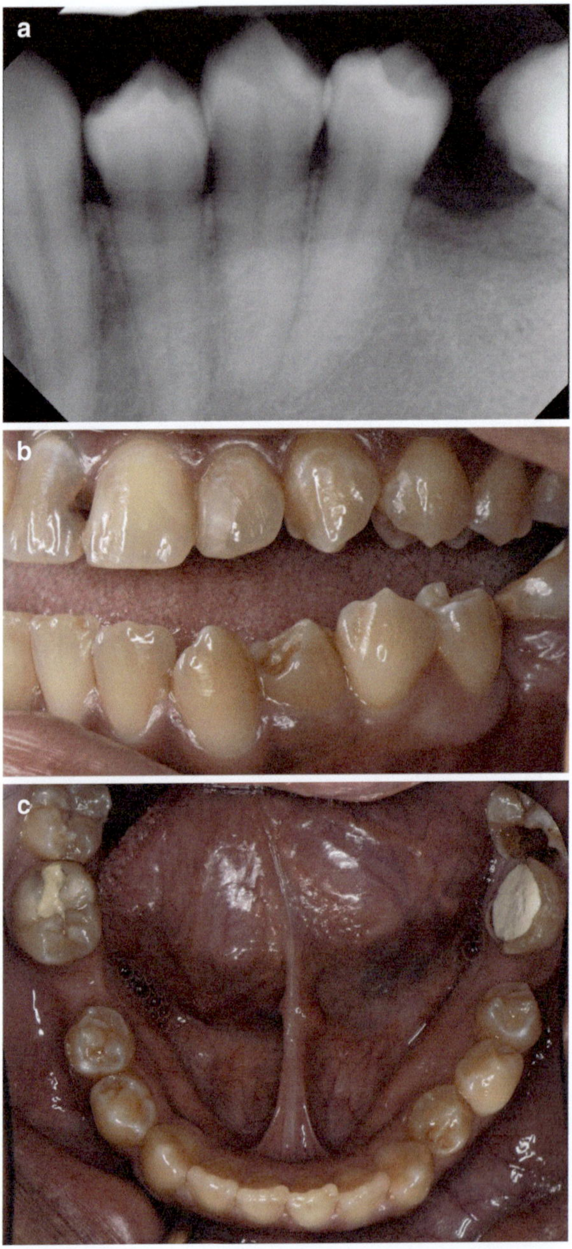

Fig. 19.3 Presence of a mandibular supplemental premolar (Courtesy of Dr. Mailis Soler)

et al. 2008; Suginami 2007). Cadenat et al. (1977) thought that an autosomal recessive gene and a gene on the inhibiting chromosome X were implicated. An autosomal dominant trait with lack of penetrance in some generations or sex-linked inheritance could explain the predominance of males over females. Abnormal reaction to a local traumatic episode and environmental factors are also proposed explanations for hyperdontia.

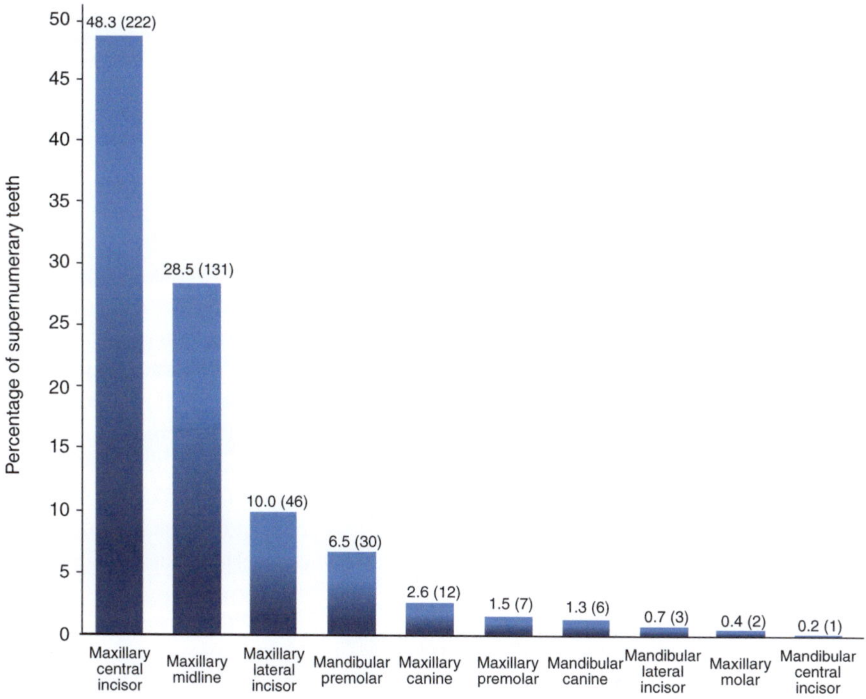

Fig. 19.4 Location of supernumerary teeth (from Gomes et al. 2008)

Table 19.1 Relation between eruption and morphology of supernumerary teeth (ST) (from Gomes et al. 2008)

	ST morphology			
	Conical	Tuberculate	Supplemental	Total
Erupted	57 (27.8%)	21 (11.8%)	29 (37.7%)	107 (23.2%)
Unerupted	148 (72.2%)	157 (88.2%)	48 (62.3%)	353 (76.8%)
Total	205 (100%)	178 (100%)	77 (100%)	460 (100%)

*Supplemental ST were more frequently erupted than conical and tuberculate ST (P < 0.001). Chi-squared test = 24.43, d.f = 2, $P < 0.001$

19.2.4 Intraoral Features with Hyperdontia

Supernumerary teeth have various shapes such as conical, tuberculate, and supplemental. Generally, supernumerary teeth are conical in the permanent dentition. In Brazilian children and adolescents, the morphology was conical in 44.6% of cases, followed by tuberculate in 38.7%, and supplemental supernumerary teeth in 16.7% (Gomes et al. 2008). Unerupted supernumerary teeth were frequently found, especially tuberculate ones, and often associated with delayed eruption of the incisors. On the other hand, conical-shaped teeth frequently erupt and do not delay the eruption of the adjacent permanent teeth (Rajab and Hamdan 2002; Gomes et al. 2008). Table 19.1 shows that 76.8% of supernumerary teeth are unerupted. Sometimes those unerupted supernumeraries impair the eruption of another tooth. The teeth adjacent to supernumeraries can be changed in shape. In terms of tooth size, patients with supernumerary teeth can be expected to have larger teeth than average (Khalaf et al. 2005) (Table 19.2).

19.2.5 Clinical Management of Hyperdontia

Although supernumerary teeth can remain impacted for many years without clinical, pathological, or orthodontic complications, in some cases they cause clinical complications such as impaction, delayed eruption, ectopic eruption overcrowding, spacing anomalies, and the formation of follicular cysts. Early diagnosis and treatment of patients with supernumerary teeth are important for preventing or minimizing such complications. However, definitive management of hyperdontia patients remains controversial, whether to remove such teeth or to monitor them. For example, some authors prefer to postpone surgical intervention for maxillary anterior supernumerary teeth until the patient reaches 8–10 years old, when root development in the central and lateral incisors is nearly complete (Henry et al. 1989), but

Table 19.2 Tooth area measurement from the occlusal view of maxillary and mandibular teeth of supernumerary and control groups (from Khalaf et al. 2005)

		Supernumerary			Control		
Tooth		*N*	Mean (mm^2)	S.D. (mm^2)	*N*	Mean (mm^2)	S.D. (mm^2)
Maxilla							
I1	M	29	50.72	6.03	20	46.14	5.05
	F	19	48.48	9.00	20	48.02	5.96
I2	M	27	37.31	5.25	16	33.68	6.22
	F	18	36.70	7.83	18	35.94	4.68
C	M	10	49.50	4.35	11	47.08	4.50
	F	7	46.57	7.32	13	46.44	3.98
P1	M	18	58.98	8.80	15	53.64	3.75
	F	18	55.33	7.11	18	53.68	4.94
P2	M	14	57.80	8.32	18	54.36	4.79
	F	14	55.97	10.11	18	55.55	6.33
M1	M	34	111.48	11.98	20	107.42	5.74
	F	21	105.33	15.16	29	100.50	9.23
	M	9	99.10	9.01	7	97.80	10.16
Mandible							
I1	VI	35	26.02	3.52	18	23.50	2.73
	F	21	25.08	4.37	20	24.85	2.12
I2	M	29	30.11	3.85	19	26.93	3.62
	F	21	28.37	4.35	20	28.34	2.23
C	M	18	40.72	5.08	14	36.41	2.93
	F	15	37.43	5.74	13	36.22	3.00
P1	M	20	49.07	8.57	15	43.79	2.77
	F	18	44.71	5.97	18	44.49	4.51
P2	M	14	52.00	5.88	15	48.99	3.46
	F	15	50.38	6.72	16	49.17	5.35
M1	M	34	103.00	12.47	20	100.33	6.90
	F	19	97.37	8.96	19	96.92	8.86
	M	4	96.92	10.36	4	91.09	5.52

others recommend extraction of mesiodens in the early mixed dentition in order to facilitate spontaneous eruption and alignment of the incisors (Tay et al. 1984). The tuberculate types of supernumerary teeth more frequently cause delayed eruption, but the conical types more frequently cause displacement of the adjacent dentition. The earlier the problematic supernumerary tooth can be removed, the better the prognosis. Supernumeraries can also lead to dentigerous cyst formation, pericoronal space ossification, and crown resorption. It is important in detecting a supernumerary tooth to rule out the presence of odontoma because the morphological characteristics of a composed odontoma are similar to those of supernumerary teeth. Both clinical and radiographic examinations are important for hyperdontia management.

The routine use of CBCT is also recommended for accurate evaluation of hyperdontia. Kara et al. (2012) showed that 27% of supernumerary molars in a Turkish population had caused at least one complication such as cyst formation, pericoronitis, diffuse dental caries, or soft tissue irritation. Analysis of supernumerary molar management approaches showed that 70% of the patients had undergone surgical removal of the teeth. Supernumerary teeth do not always need to be extracted. Unerupted supernumerary teeth that are symptomless (i.e., do not appear to be affecting the dentition) are sometimes left alone under careful observation. A routine followup visit is recommended for those patients. Hyperdontia frequently occurs in persons with other dental anomalies and developmental disorders. Patients with cleft lip and palate are at greater risk for hyperdontia, reports indicating an incidence of up to 4.9% in Chinese children (Fan et al. 2011). Classic syndromes involving hyperdontia can include Gardner syndrome, cleidocranial dysplasia, Fabry-Anderson syndrome, and Ehlers-Danlos syndrome; cleidocranial dysplasia has the strongest association with this dental anomaly.

19.3 Congenitally Missing Teeth (Hypodontia)

19.3.1 Definition and Classification

Congenitally missing teeth (CMT) is classified on the basis of the number of absent teeth. The absence of one to five teeth is termed hypodontia, while the absence of more than six teeth is termed oligodontia; congenital absence of all teeth is anodontia. Anodontia is rare and mostly occurs in cases of ectodermal dysplasia. In general, CMT is more prevalent than hyperdontia.

19.3.2 Prevalence

19.3.2.1 In Primary Dentition

CMT is rather uncommon in the primary dentition (prevalence 0.1–0.9%) and affects both genders equally (Dhanrajani and Hypodontia 2002; Nunn et al. 2003). It usually affects the anterior maxilla, particularly the lateral incisors (Daugaard-Jensen et al. 1997). There is a strong correlation between CMT in primary and permanent dentitions (Bailleul-Forestier et al. 2008). There are reports of children with primary CMT showing absence of the corresponding successor teeth (Bailleul-Forestier et al. 2008; Nik-Hussein et al. 1996; Whittington et al. 1996).

19.3.2.2 In Permanent Dentition

A meta-analysis by Polder et al. (2004) reported the average prevalences of CMT in Europe (5.5%), North America (3.9%), and China (6.9%) (Table 19.3). These authors also investigated the sites and frequency of missing teeth in the permanent dentition, reviewing 10 studies with an aggregate sample of over 48,000 people. The frequencies of absent teeth (Table 19.4) were:

Mandibular second premolar (3.0%)
Maxillary lateral incisor (1.7%)
Maxillary second premolar (1.5%)
Mandibular central incisor (0.3%)
Mandibular lateral incisor and maxillary first premolar (0.2%)
Mandibular first premolar (0.15%)
Mandibular second molar and maxillary canine (0.1%)
Maxillary second molar (0.05%)
Maxillary first molar (0.03%)
Mandibular canine (0.02%)
Mandibular first molar (0.01%)
Maxillary central incisor (0.005%)

Table 19.3 Prevalence of dental agenesis by continent, race and gender in percentages (and 95% CI) (from Polder et al. 2004)

	Males	Females	Total
Europe (Caucasian)	4.6 (4.5–4.8)	6.3 (6.1–6.5)	5.5 (5.3–5.6)
North America (Caucasian)	3.2 (2.9–3.5)	4.6 (4.2–4.9)	3.9 (3.7–4.1)
North America (African American)	3.2(2.2–4.1)	4.6 (3.5–5.8)	3.9 (3.1–4.6)
Australia (Caucasian)	5.5 (4.4–6.6)	7.6 (6.0–9.2)	6.3 (5.4–7.2)
Saudi Arabia (Caucasian)	2.7 (2.0–3.4)	2.2 (1.2–3.1)	2.5 (1.9–3.1)
Chinese (Mongoloid)	6.1 (4.0–8.1)	7.7 (5.4–10.0)	6.9 (5.3–8.4)

Table 19.4 Prevalence in percentages and 95% CI of dental agenesis of individual teeth derived from 48,274 persons (from Polder et al. 2004)

	Maxilla		Mandible	
	n	Prevalence (95% CI)	*n*	Prevalence (95% CI)
I1	3	0.00–0.01	143	0.25–0.35
I2	804	1.55–1.78	102	0.17–0.25
c	47	0.07–0.25	8	0.01–0.03
P1	100	0.170.25	66	0.10–0.17
P2	722	1.39–1.61	1479	2.91–3.22
M1	17	0.02–0.05	6	0.00–0.02
M2	21	0.03–0.06	47	0.07–0.13

Most (83%) patients with developmental hypodontia had only one or two teeth missing. Patients lacking three to five teeth represented 14.4% of the group, while 2.6% had six or more absent teeth (Polder et al. 2004) (Figs. 19.5 and 19.6). This was equivalent to a population prevalence of 0.14%. It is more common for maxillary lateral incisors to be absent bilaterally and other teeth to be absent unilaterally (Polder et al. 2004) (Fig. 19.7). It has been hypothesized that the prevalence of hypodontia in permanent teeth has increased over the years (Mattheeuws 2004). However, Dhamo et al. (2015) compared the prevalences of hypodontia in 1970 and 2010 and found no statistically significant difference. Hypodontia was significantly more prevalent in females than males (Polder et al. 2004).

19.3.3 Etiology

The etiology of congenitally missing teeth is still not clear. It is multifactorial, involving a combination of environmental and genetic factors.

19.3.3.1 Tooth Agenesis Hypothesis

Early workers investigating the etiology of isolated non-syndromic hypodontia suggested an anthropological perspective, pointing to an ongoing process of evolution. Dahlberg (1945) suggested that the most mesial tooth in four morphological fields (incisors, canines, premolars, and molars) was the most genetically stable and seldom missing. On the other hand, teeth at the end of each field (second premolars, lateral incisors, and third molars) were less genetically stable. This led to the concept of stable and unstable elements in the dentition (Bailit 1975).

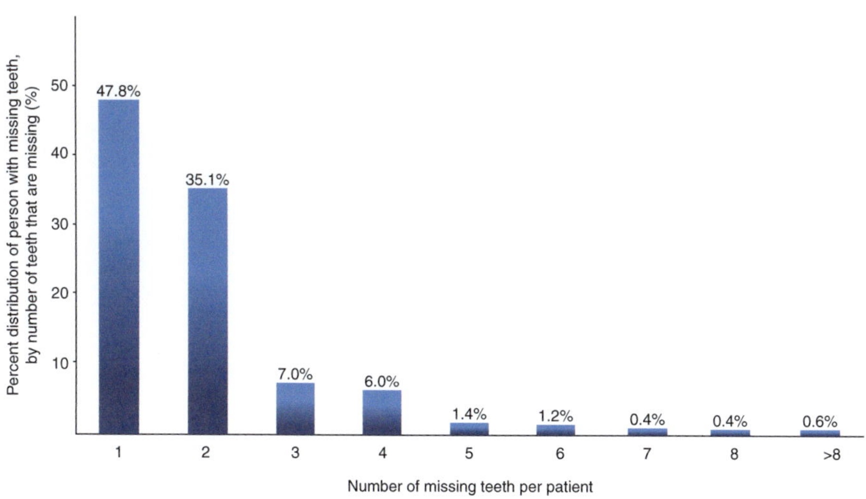

Fig. 19.5 Percentage distribution of persons with missing teeth, by number of teeth that are missing (from Polder et al. 2004)

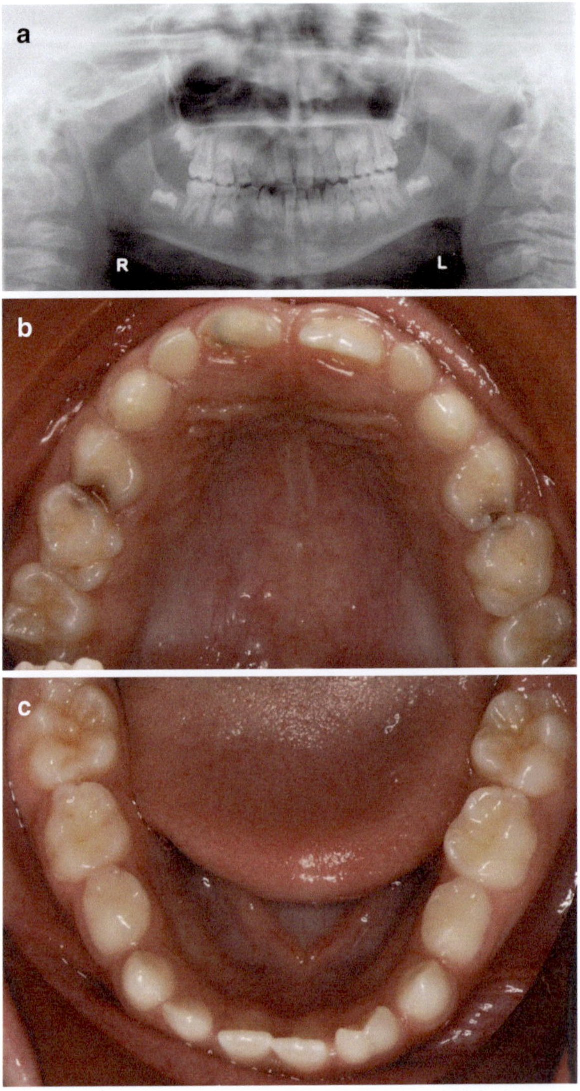

Fig. 19.6 8y5m male patient. Congenitally missing #s 1, 6, 7, 10, 16, 17, 22, 23 (N is geminated), 23, 26, 27, 28 and 32. (Courtesy of Dr. Christopher L. Connell)

19.3.3.2 Environmental Factors

Environmental factors that can cause congenitally missing teeth include developmental anomalies, endocrine disturbances, infections such as rubella, irradiation, facial trauma such as iatrogenic injury to the developing tooth germ from traumatic extraction of the overlying primary tooth, and medical treatment.

19.3.3.3 Genetic Factors

Genetic factors have been shown to be involved in hypodontia. A mutation causes selective tooth absence. Hypodontia has been associated with cleft lip and palate.

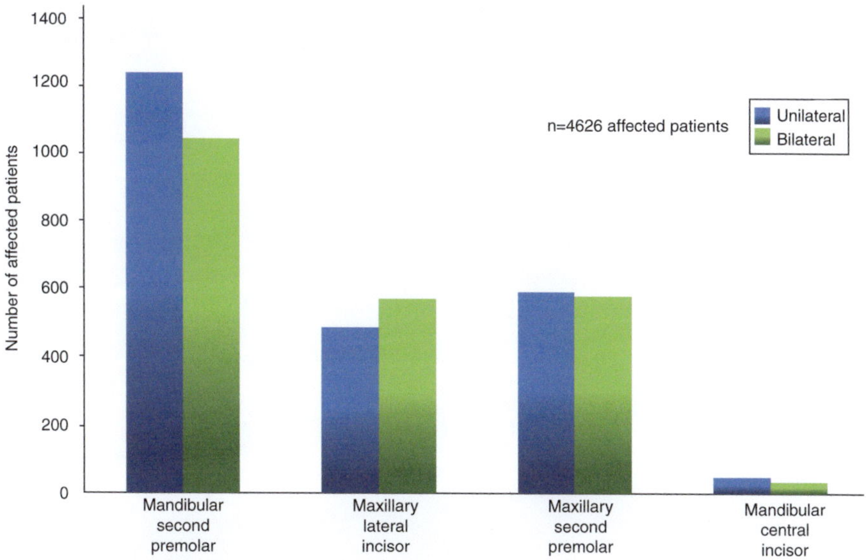

Fig. 19.7 Unilateral and bilateral occurrence of agenesis for the four most affected teeth (from Polder et al. 2004)

Among complete unilateral cleft lip and palate patients in a population of Chinese children, 66.8% presented with hypodontia (Fan et al. 2011). This was recently considered to be a defect in the Msx1 gene, which is associated with both isolated cleft lip and cleft palate and with hypodontia (Satokata and Maas 1994; Van den Boogaard et al. 2000; Alappat et al. 2003). Hypodontia can also occur in hereditary syndromes such as Crouzon syndrome, Down's syndrome, ectodermal dysplasia, Hurler syndrome, and Turner syndrome.

19.3.4 Intraoral Features with Hypodontia

Individuals with hypodontia have characteristic skeletal features and growth patterns. There are several clinical features including microdontia, delayed eruption of other teeth, retained primary teeth, transposition of permanent teeth, ectopic permanent teeth, abnormal dental morphologies such as taurodontism and peg-shaped maxillary lateral incisors, and infraocclusion of primary molar teeth.

Hypodontia and microdontia can occur together in part of the mouth or the whole dentition. Brook's study (1984) revealed a correlation between hypodontia and microdontia. This can be seen in both the primary and permanent dentitions and can affect one or more teeth, and abnormal crowns and contours can also be seen. Microdontia often occurs at the maxillary lateral incisor (peg lateral). Patients with ectodermal dysplasia can have the severest form of microdontia and the many other syndromes linked with it (Fig. 19.8).

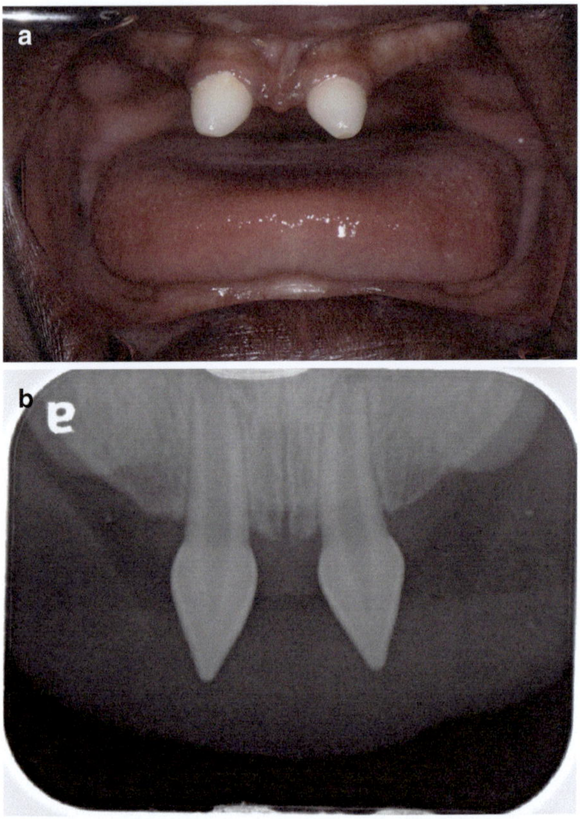

Fig. 19.8 Ectodermal dysplasia with conical shaped teeth (Courtesy of Dr. Caroline Truong)

Ectopic positioning of the permanent teeth is another feature of hypodontia. Ectopic eruption of maxillary canines is well documented. It probably occurs because of missing or small maxillary lateral incisors, missing second premolars, retained primary canines, impacted permanent canines and central incisors, and severe rotation of the adjacent premolars.

Delays in tooth development are also common. Exfoliation of primary teeth is mainly caused by resorption of their roots due to eruption of the permanent successors. However, absence of permanent successors delays the normal resorption process. Haselden et al. (2008) reported that some primary teeth without permanent successors can persist into the 50s and in some cases are retained in the 60s. Those long-term retained primary teeth were most likely to be canines and a few molars. This has important implications for the time of treatment planning for hypodontia patients. The canines can retain their function for many years, but molars, especially the first molars, are not as reliable.

Altered craniofacial morphology is another feature of CMT patients. The short face height and large freeway space typical of hypodontia patients can make them appear over-closed, more intensely if the severity of missing teeth is greater (Hobkirk et al. 2011). In a cephalometric study by Chung et al. (2000), the mean values for the sample as a whole were within the standard range and showed no features typical of the hypodontia group. However, patients with more severe congenitally missing teeth (when more than one tooth type was missing) tended to have a reduced maxillary-mandibular plane angle and a class III skeletal relationship. Acharya et al. (2010) found that the A-P skeletal relationship tended toward a Class III pattern with increasing severity of hypodontia. This was clinically significant only when more than six teeth were missing. Kumar et al. (2013) found that the pattern of anterior missing teeth showed a shorter posterior cranial base, a significantly more retrognathic maxilla and prognathic mandible, flat chin, decreased upper anterior facial height, decreased mandibular angle, and elongation of the upper incisors, which could be a functional compensation for the retrognathic maxilla.

19.3.5 Clinical Management

Different CMT patients have different problems, so treatment in each case should be based on its pros and cons and communicated among the interdisciplinary team. Pediatric dentists, prosthodontists, restorative dentists, orthodontists, and oral surgeons should be involved. Age is important for determining the type of treatment for CMT. Depending on the patient's concerns, treatment should be kept to a minimum for the primary and mixed dentition. Restorative intervention can sometimes be avoided at this stage. CMT needs to be treated with dentures as early as 3 years of age, and dental implants are a treatment option for older individuals. Bonding composite for conical-shaped teeth also helps to improve cosmetic appearance and function. Orthodontic treatment can help with any necessary restorative treatment. Common problems in treating hypodontia patients include space management, uprighting and aligning teeth, management of the deep overbite, and retention. The amount of spacing is influenced by microdontia, retention of the primary teeth, and the abnormal eruptive paths and drifting of the successional teeth (Hobkirk et al. 2011). The decision about whether the treatment plan involves space closure or opening the spaces of the missing mandibular second premolar depends on factors such as the age of the patient, the degree of inherent crowding, the state of the deciduous teeth, the type of malocclusion, and the circumstances of the patient (finances, attitude toward treatment, etc.).

Depending on patient age, the following are examples of treatment for CMT (Sadaqah and Tair 2015):

For patients younger than 6 years [primary dentition] with severe hypodontia:

- Removable partial, which can help psychologically and functionally

For patients in the age range 7–12 years [mixed dentition], one or a combination of the following options can be chosen

- Removable partial denture
- Composite buildups for microdont permanent teeth to improve cosmetic appearance and function

- Extractions to guide eruption of permanent successors
- Orthodontic treatment.

F*or patients in the age range 12–16 years [permanent dentition], one or a combination of the following options can be chosen:*

- Orthodontic treatment
- Resin-bonded bridges following orthodontic treatment
- Composite buildups for microdont permanent teeth to improve cosmetics and function
- Overdentures for severe hypodontia

For patients aged between 16 and 20, one or combination of the following treatment options can be chosen:

- Single-tooth implant or implant-fixed bridges or implant-retained overdentures after the jaw bone has finished growing
- Orthodontic in combination with orthognathic surgery

References

Acharya PN, Jones SP, Moles D et al (2010) A cephalometric study to investigate the skeletal relationships in patients with increasing severity of hypodontia. Angle Orthod 80:699–706

Acıkgoz A, Acıkgoz G, Tunga U et al (2006) Characteristics and prevalence of non-syndrome multiple supernumerary teeth: a retrospective study. Dentomaxillofac Radiol 35:185–190

Alappat S, Zhang ZY, Chen YP (2003) MSX homeobox gene family and craniofacial development. Cell Res 13:429–442

Bailit HL (1975) Dental variation among populations. An anthropologic view. Dent Clin N Am 19:125–139

Bailleul-Forestier I, Molla M, Verloes A et al (2008) The genetic basis of inherited anomalies of the teeth. Part 1: clinical and molecular aspects of non-syndromic dental disorders. Eur J Med Genet 51:273–291

Brook AH (1984) A unifying aetiological explanation for anomalies of human tooth number and size. Arch Oral Biol 29:373–378

Cadena H, Combelles R, Fabert G et al (1977) Mesiodens and heredity (Mesiodens et heredite). Rev Stomatol Chir Maxillofac 78:341–346

Chung LK, Hobson RS, Nunn JH et al (2000) An analysis of the skeletal relationships in a group of young people with hypodontia. J Orthod 27:315–318

Dahlberg AA (1945) The changing dentition of man. J Am Dent Assoc 32:676–690

Daugaard-Jensen J, Nodal M, Skovgaard LT et al (1997) Comparison of the pattern of agenesis in the primary and permanent dentitions in a population characterized by agenesis in the primary dentition. Int J Paediatr Dent 7:143–148

Dhamo B, Vucic S, Kuijpers MA et al (2015) The association between hypodontia and dental development. Clin Oral Investig 20:1347–1354

Dhanrajani PJ (2002) Hypodontia DP. Etiology, clinical features, and management. Quintessence Int 33:294–302

Fan XX, Li J, Ge LH et al (2011) Dental anomalies in Chinese children with complete unilateral cleft lip and palate. Zhonghua Kou Qiang Yi Xue Za Zhi 46:263–266

Garvey MT, Barry HJ, Blake M (1999) Supernumerary teeth – an overview of classification, diagnosis and management. J Can Dent Assoc 65:612–616
De Oliveira Gomes C, Drummond SN, Jham BC et al (2008) A survey of 460 supernumerary teeth in Brazilian children and adolescents. Int J Paediatr Dent 18:98–106
Haselden K, Hobkirk JA, Goodman JR et al (2008) Root resorption in retained deciduous canine and molar teeth without permanent successors in patients with severe hypodontia. Int J Paediatr Dent 11:171–178
Henry RJ, Post AC (1989) A labially positioned mesiodens: case report. Pediatr Dent 11:59–63
Hobkirk JA, Gill DS, Jones SP et al (2011) Hypodontia a team approach to management. Wiley-Blackwell, Hoboken
Kara Mİ, Aktan AM, Ay S et al (2012) Characteristics of 351 supernumerary molar teeth in Turkish population. Med Oral Patol Oral Cir Bucal 17:e395–e400
Khalaf K, Robinson DL, Elcock C et al (2005) Tooth size in patients with supernumerary teeth and a control group measured by image analysis system. Arch Oral Biol 50:243–248
Kumar SK, Lakshmi AV, Namita S et al (2013) Craniofacial morphologic variations and its association with hypodontia pattern (anterior) in South Indian female population. Biosci Biotechnol Res Asia 10:325–328
Mahabob MN, Anbuselvan GJ, Kumar BS et al (2012) Prevalence rate of supernumerary teeth among non-syndromic south Indian population: an analysis. J Pharm Bioallied Sci 4:373
Mattheeuws N (2004) Has hypodontia increased in caucasians during the 20th century? A meta-analysis. Eur J Orthod 26:99–103
Nik-Hussein NN, Abdul Majid Z (1996) Dental anomalies in the primary dentition: distribution and correlation with the permanent dentition. J Clin Pediatr Dent 21:15–19
Nunn JH, Carter NE, Gillgrass TJ et al (2003) The interdisciplinary management of hypodontia: background and role of paediatric dentistry. Br Dent J 194:245–251
Polder BJ, Van't Hof MA, Van der Linden FP et al (2004) A meta-analysis of the prevalence of dental agenesis of permanent teeth. Community Dent Oral Epidemiol 32:217–226
Primosch RE (1981) Anterior supernumerary teeth – assessment and surgical intervention in children. Pediatr Dent 3:204–215
Rajab LD, Hamdan MAM (2002) Supernumerary teeth: review of the literature and a survey of 152 cases. Int J Paediatr Dent 12:244–254
Sadaqah NR, Tair JA (2015) Management of patient with hypodontia: review of literature and case report. Open J Stomatol 5:293–308
Satokata I, Maas R (1994) Msx1 deficient mice exhibit cleft palate and abnormalities of craniofacial and tooth development. Nat Genet 6:348–356
Suginami A (2007) Poster 227: rudiment incisors survive and erupt as supernumerary teeth as a result of USAG-1 abrogation. J Oral Maxillofac Surg 65:43.e126
Tay F, Pang A, Yuen S (1984) Unerupted maxillary anterior supernumerary teeth: report of 204 cases. ASDC J Dent Child 51:289–294
Taylor GS (1972) Characteristics of supernumerary teeth in the primary and permanent dentition. Dent Pract Dent Rec 22:203–208
van den Boogaard MJ, Dorland M, Beemer FA et al (2000) Erratum to MSX1 mutation is associated with orofacial clefting and tooth agenesis in humans. Nat Genet 25:342–343
Whittington BR, Durward CS (1996) Survey of anomalies in primary teeth and their correlation with the permanent dentition. N Z Dent J 92:4–8

Variations in the Anatomy of the Teeth 20

Yasuhiko Kamura

20.1 Introduction

For the success of dental therapy, knowledge of tooth anatomy cannot be ignored. It is essential to know the normal configurations of the teeth and their variations. There are numerous variations in the internal anatomy of teeth due to developmental anomalies, hereditary factors, trauma, etc.

20.1.1 Variations in Development

1. Amelogenesis Imperfecta

   Amelogenesis imperfecta (AI) comprises a group of hereditary disorders of dental enamel. The reported incidences range between 43:10,000 in Turkey (Altug-Atac and Erdem 2007), 14:10,000 in Sweden (Bäckman and Holm 1986), 10:10,000 in Argentina (Sedano 1975), and 1.25:10,000 in Israel (Chosack et al. 1979). The structure and clinical appearance of the enamel are affected in both primary and permanent dentition, and there is hypomineralization and/or hypoplasia with discoloration, sensitivity, and fragility. The main aim of treatment should be early diagnosis, pain management, prevention, stabilization, restoration of defects, and regular maintenance (Fig. 20.1).
2. Enamel Pearl

   Ectopic globules of enamel, or so-called enamel (enamelous) pearls, droplets, globules, nodules, knots, or exostoses, are a developmental anomaly of teeth (Moskow and Canut 1990). They can be either internal or external, the former being more common on the root surface. Internal ectopic globules are

Y. Kamura (✉)
Sigma Dental Specialists, Murphy, TX, USA

J. Iwanaga, R. S. Tubbs (eds.), *Anatomical Variations in Clinical Dentistry*,
https://doi.org/10.1007/978-3-319-97961-8_20

referred to as intradental, interdental, or intradentinal enamel pearls. The mean overall prevalence of enamel pearls is 2.69%, but the prevalence for molars is 2.28% (Moskow and Canut 1990). Clinically, enamel pearl is a risk for localized deep periodontal pockets since it compromises the attachment of the periodontal ligament.

3. Enamel Projection

   Cervical enamel projection is a developmental abnormality causing focal triangular extension of the enamel beyond the normal CEJ to the furcation area. Cervical enamel projection can be considered a secondary etiological factor for periodontal breakdown and attachment loss due to the lack of connective tissue attachment (Swan and Hurt 1976). Clinically, removal of ectopic enamel by periodontal surgeries is generally recommended to allow the connective tissue attachment to form (Fig. 20.2).

4. Gemination

   Gemination occurs when a single tooth bud divides into two separate entities. This results in a structure with two completely or incompletely separated crowns with a single root and root canal. Consequently, the patient has a bigger

Fig. 20.1 Amelogenesis imperfecta (Courtesy of Dr. Teppei Tsukiyama)

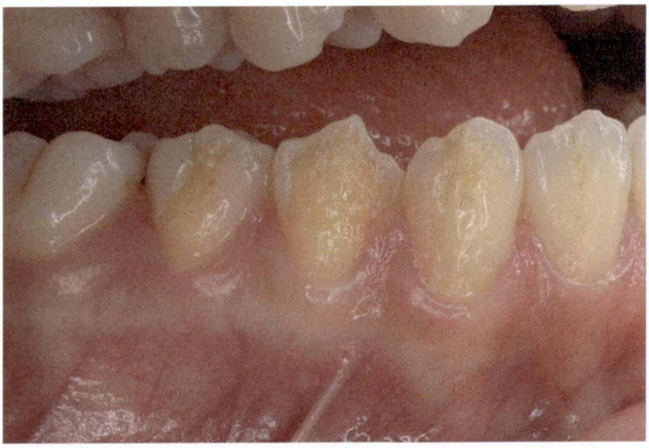

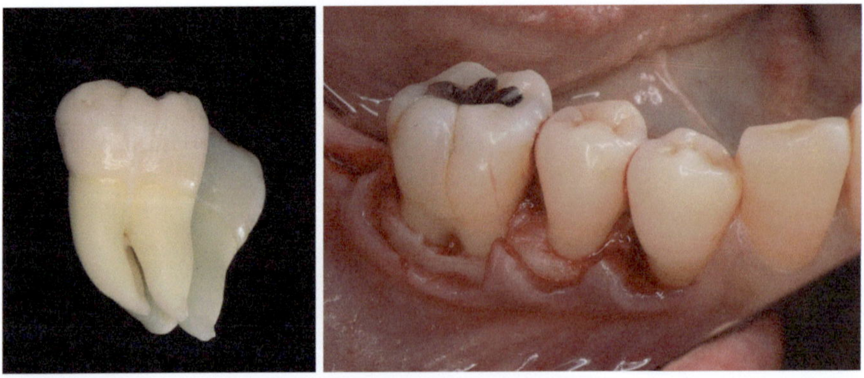

Fig. 20.2 Enamel projection (Courtesy of Dr. Teppei Tsukiyama)

tooth, but the number of teeth in the affected arch is normal when the geminated tooth is counted as one. This condition can also cause impaction because the crown of the geminated tooth is bigger than normal. Gemination in the anterior maxillary region is commonly documented in the literature (Dexton et al. 2011). The etiology is unknown, but trauma and familial tendency have been proposed (Järvinen et al. 1980). Gemination is observed in both deciduous and permanent dentitions (O Carroll 1990). Clinically, fissures or grooves in the union between the teeth are a risk for caries and periodontal disease. Also, owing to the abnormal morphology of the crown and the complexity of the root canal system in geminated teeth, root canal treatment protocols require special attention. CBCT is advocated to help the clinician understand the anatomy (Matherne et al. 2008) (Fig. 20.3).

5. Fusion

   Fusion of teeth is a developmental anomaly that results in the union of two normally separated tooth germs. Fused teeth can have separate or fused pulp spaces (Neville et al. 1993). The incidence ranges between 0.3% and 3.8% in the general population (Yusof 1990). The most important factor distinguishing this condition from germination is that fusion usually leads to a reduced number of teeth in the dentition. As with germination, the anomaly entails a high risk for dental caries, periodontal disease, and spacing between teeth (Fig. 20.3).

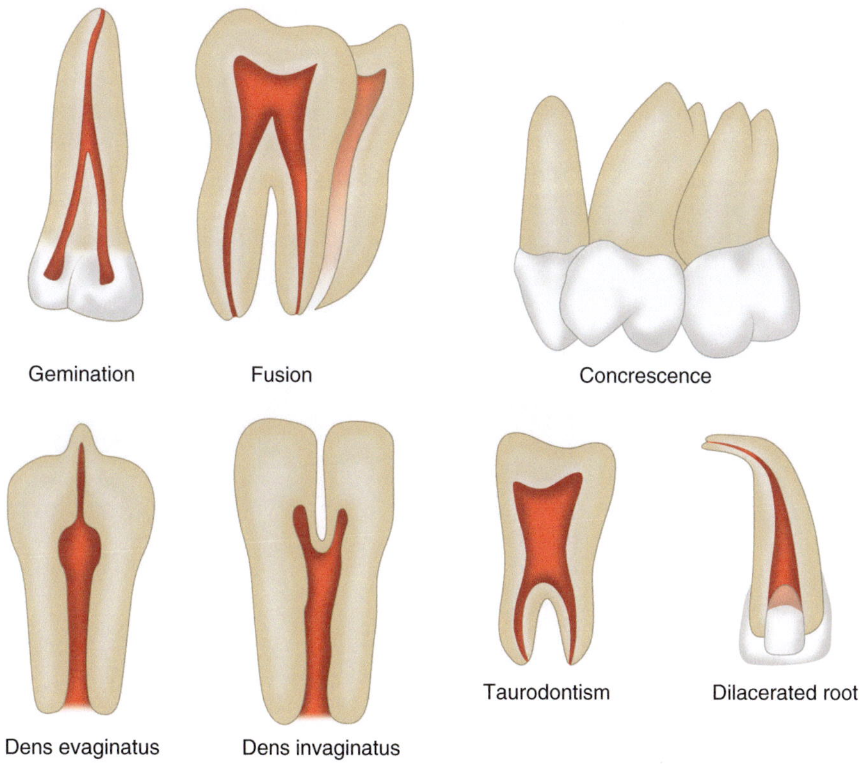

Fig. 20.3 Schematic representation of tooth anomalies

6. Concrescence

   Concrescence is an uncommon developmental anomaly that occurs after root formation is complete: there is a confluence of the cementum of two teeth at the root level but not the dentine. The reported frequency of concrescence is 0.8% in adult teeth and 0.2–3.7% in deciduous teeth, and the incidence is highest in the posterior maxilla (Law et al. 1994). Some reported complications of concrescence include periodontal problems (Mohan 2014). Owing to their large mesiodistal dimensions, extractions of concrescent teeth can be difficult and can result in alveolar bone fracture and tooth fracture, or in sinus opening (Fig. 20.3).

7. Taurodontism

   Taurodontism is a morpho-anatomical change in tooth shape in which the body of the tooth is enlarged and the roots are small. The pulp chamber of this tooth is extremely large with a greater apico-occlusal height than normal. From a periodontal standpoint, taurodont teeth can offer a favorable prognosis in specific cases since there can be furcations only few millimeters above the root apex. From an endodontist's view, taurodontism presents a challenge during negotiation, instrumentation, and obturation in root canal therapy as the pulp lacks the normal constriction at the cervical level of the tooth. Taurodontism is often associated with syndromes such as Klinefelter and Down syndromes. Its etiology is commonly attributed to the failure of invagination of the epithelial root sheath sufficiently early to form the cynodont (Witkop 1971) (Figs. 20.3 and 20.4).

8. Talon Cusp

   Talon cusp is a developmental anomaly in which an accessory cusp-like or tooth-like structure forms from the cingulum area or cementoenamel junction to the incisal edge of the maxillary or mandibular anterior teeth. The reported prevalence is between 0.06% and 7.7% (Lee et al. 2006). The etiology is not completely understood. Clinically, we have to be very careful in treating talon cusps since the condition can cause esthetic, caries, and periodontal problems.

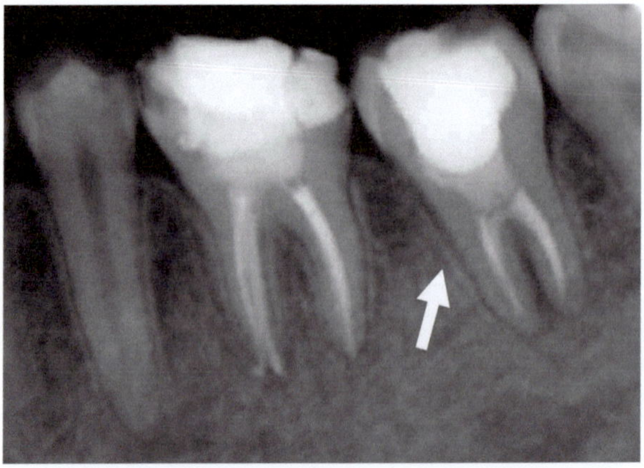

Fig. 20.4 Taurodont tooth #18 (arrow) (Courtesy of Dr. Atsushi Fujimura)

9. Dilaceration

   Dilaceration is the result of a developmental anomaly leading to an extraordinarily sharp bend in the roots of the teeth. Two possible etiologies of dilaceration are related to trauma and to disturbances during root development. Dilaceration can be seen in both the permanent and deciduous dentitions, and it is more commonly found in the posterior teeth and in the maxilla. Radiographic information from periapical radiographs or CBCT facilitates the correct diagnosis of root dilacerations. The excessive root curvature obviously complicates dental treatment especially endodontic procedures (Fig. 20.3).

10. Dentinogenesis Imperfecta

    Dentinogenesis imperfecta is the most common type of developmental anomaly of dentine. Dentitions are affected by an opalescent brown discoloration, and because of the reduced support of the dentine, the overlying enamel fractures readily. There is also rapid wear and attrition of the teeth. There are various degrees of partial or total precocious pulp obliteration. As dentinogenesis imperfecta is associated with rapid wear, attrition of the teeth, and crown fractures, protection from tooth wear soon after eruption is strongly recommended (Seow and Taji 2011) (Fig. 20.5).

11. Dentine Dysplasia

    Clinically, the crowns of the affected teeth are usually normal in shape and color but sometimes slightly opalescent. Radiographically, the pulp chambers are largely or completely obliterated and the roots are short, often pointed with apical radiolucencies in the absence of caries.

12. Palatogingival groove

    Palatogingival groove (PGG) is an anomaly in the maxillary anterior teeth, often with a deep narrow periodontal pocket that communicates with the pulp causing endodontic-periodontal lesions. PGG can be classified by its location, extent, complexity, etc. The groove acts as a trap that can harbor plaque and bacteria. Three strategies are advocated for treating teeth with PGG: (1) microorganism elimination, (2) permanent sealing of the root groove that communicates between the root canal system and the periodontium, and (3)

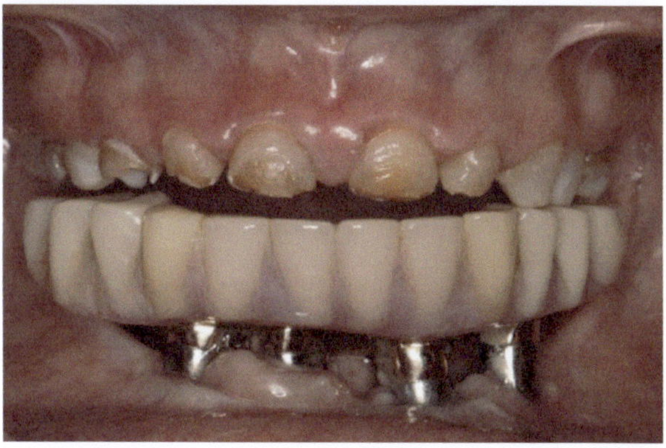

Fig. 20.5 Dentinogenesis imperfecta (Courtesy of Dr. Teppei Tsukiyama)

periodontal regeneration and complete healing of the periodontium. If the dental pulp is involved, endodontic treatment is recommended before the periodontal approach (Fig. 20.6).

13. Dens in Dente or Dens Invaginatus

    Dens invaginatus is a developmental anomaly involving invagination of the enamel organ into the dental papilla during morphodifferentiation, which results in a small tooth within the pulp chamber. Types of dens invaginatus can be classified as follows, according to Oehler (1957) (Fig. 20.7):

    - Type I: The invagination is minor and enamel-lined; it is confined within the crown of the tooth and does not extend beyond the level of the external enamel-cement junction.
    - Type II: The invagination is enamel-lined and extends into the pulp chamber but remains within the root canal with no communication with the periodontal ligament.
    - Type III A: The invagination extends through the root and communicates laterally with the periodontal ligament space through a pseudo-foramen.

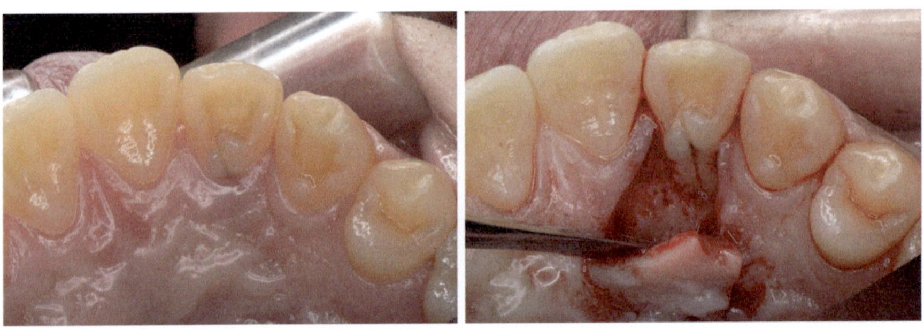

Fig. 20.6 Palatogingival groove (Courtesy of Dr. Teppei Tsukiyama)

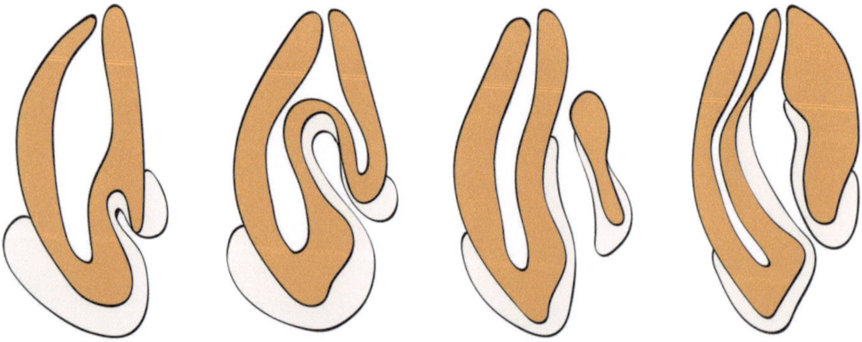

Fig. 20.7 Oehler's classification of dens invaginatus (coronal types) cited from Oehlers (1957)

There is usually no communication with the pulp, which lies compressed within the root.

- Type III B: The invagination extends through the root and communicates with the periodontal ligament at the apical foramen. There is usually no communication with the pulp.

The condition is most commonly found in the maxillary lateral incisors. It predisposes the tooth to early decay and therefore pulpitis. Conventional radiography in the classification and management of dens invaginatus is being superseded by the increasing availability of CBCT imaging (Ball et al. 2013). In Type III lesions, any infection within the invagination can lead to inflammation within the periodontal tissues giving rise to a "peri-invagination periodontitis." The treatment options are prophylactic or preventive sealing of the invagination, root canal treatment, endodontic apical surgery, intentional replantation, and extraction. The prognosis following any treatment is often dubious (Figs. 20.3 and 20.8).

14. Dens Evaginatus

    Dens evaginatus is most common in the mandibular premolars and in individuals of Asian ancestry. Clinically, it appears as an anomalous small tubercle

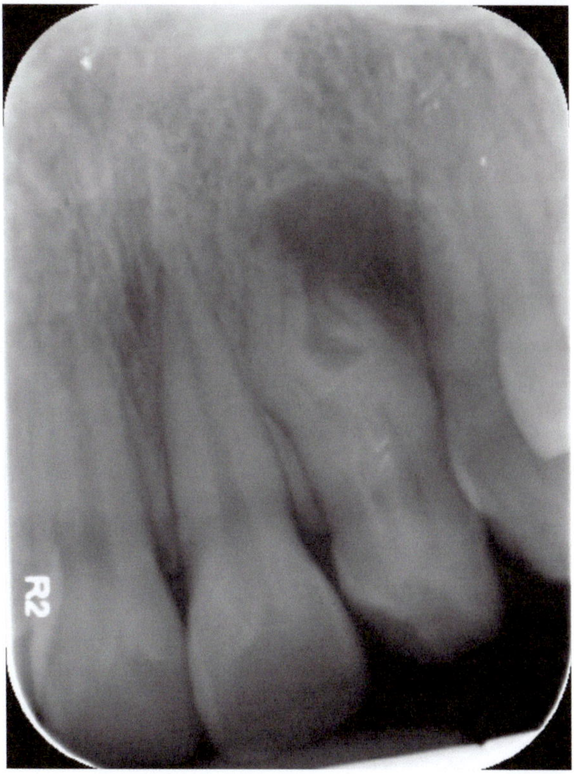

Fig. 20.8 Dens invaginatus type II tooth #10 (Courtesy of Dr. Atsushi Fujimura)

or bulge on the occlusal surface. Occlusal abrasion or tubercle fracture causes early exposure of the pulp horn that extends into the tubercle. This can further result in periapical pathology. To prevent tubercle fractures, the tubercle is removed with a burr and the tooth is capped, followed by a good sealing restoration with composite (Fig. 20.9).

20.2 Variations in Pulp Cavity Due to Pathology

20.2.1 Pulp Stones and Calcifications

Pulp stones are usually found in the coronal pulp in the form of discrete, concentric calcifications, whereas in the radicular pulp, the calcification is more likely to be

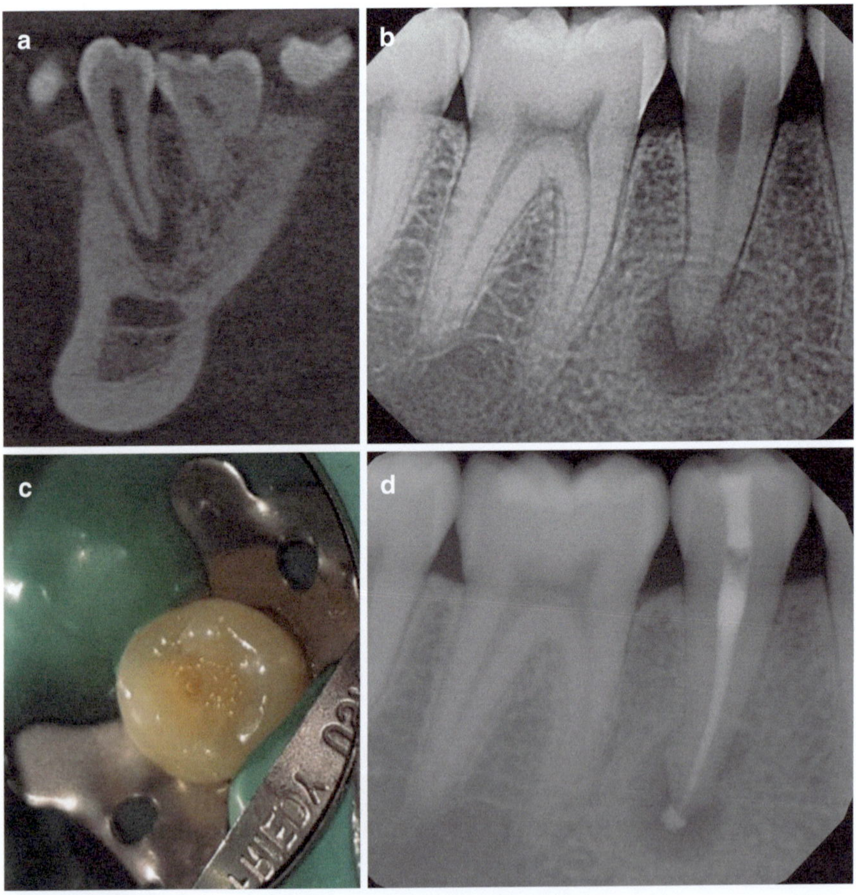

Fig. 20.9 Dens evaginatus (Courtesy of Dr. Jae Ha Jung). (**a**) Preoperative CBCT. (**b**) Preoperative radiograph. (**c**) Fractured tubercle. (**d**) Postoperative radiograph

diffuse (Sundqvist and Lerner 1996). Typically, calcification of root canals begins at the coronal aspect and extends to the apical canal. The clinical significance of pulp stones and calcifications is the obstruction of endodontic treatment. Magnification and illumination are essential for identifying and treating calcified canals. CBCT can also assist the clinician by locating the pulp stones and identifying the severity of the calcification during perioperative assessment (Kuyk and Walton 1990).

Pulp stones or denticles were once classified as true or false, depending on the presence of a tubular structure. This classification has been challenged, and a new nomenclature based on the genesis of the calcification has been suggested. Three types of pulp stones have been also classified according to location as follows: free stones, attached stones, and embedded stones.

Extensive calcification, also known as calcific metamorphosis, occurs due to trauma, caries, periodontal disease, or irritants. Palpation and percussion are usually within normal limits. This condition is not pathological and does not require treatment (Figs. 20.10 and 20.11).

20.2.2 Resorption

Resorption can be classified as internal or external. Pulp-periodontal communications due to resorptions are often complex.

Limited internal resorption can be managed without great difficulty, but external apical resorption can drastically alter the anatomy of the apex or the root surface.

1. Internal Resorption

   Internal resorption is an intracanal resorption commonly seen in the central maxillary incisors, though any tooth can be affected. It is characterized by oval enlargement of the root canal space. A possible etiology is inflammation in the pulp. The pulp is transformed into vascularized inflammatory tissue with dentinoclastic activity; this leads to the resorption of the dentinal walls (Walton and Leonard 1986). Teeth with intracanal resorptive lesions usually respond

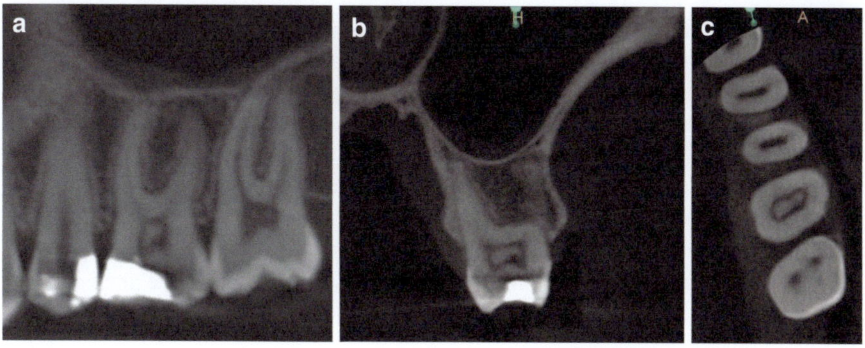

Fig. 20.10 Pulp stone. (**a**) Sagittal plane. (**b**) Frontal plane. (**c**) Axial plane

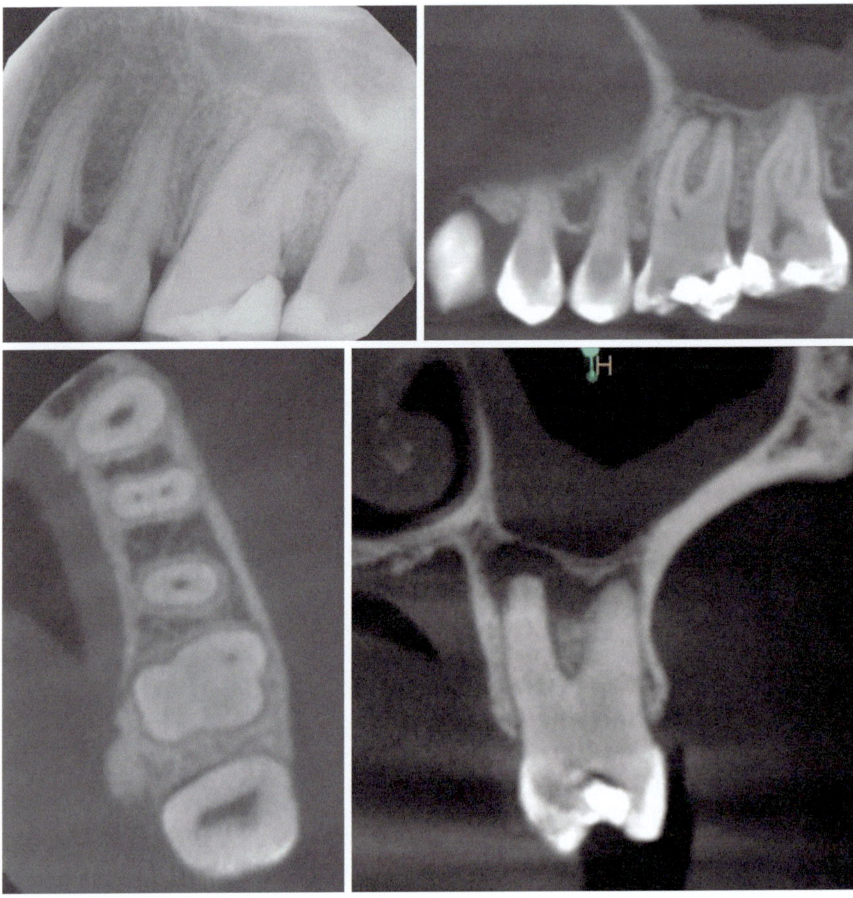

Fig. 20.11 Calcific metamorphosis

within normal limits to pulpal and periapical tests and are asymptomatic clinically. As long as there is no communication to the periodontium, the prognosis is favorable. Pink spots in the crown are often observed with advanced internal resorption involving the pulp chamber.

Radiographs especially CBCT can reveal radiolucency with irregular enlargement of the root canal anatomy. Removal of the inflamed tissue and completion of root canal treatment will arrest the resorption. If the root canal procedure is not done immediately, these lesions tend to be progressive and eventually perforate to the periodontium. Perforating internal resorptions often create operative difficulties and a guarded prognosis. With internal resorption, it is difficult to negotiate the canal from the coronal portion to the apex. Also, it is difficult to remove tissues and bacteria from the resorptive defect (Fig. 20.12).

2. External Resorption

External root resorption is initiated in the periodontium and affects the external or lateral surface of the root. It is often observed in trauma cases, especially

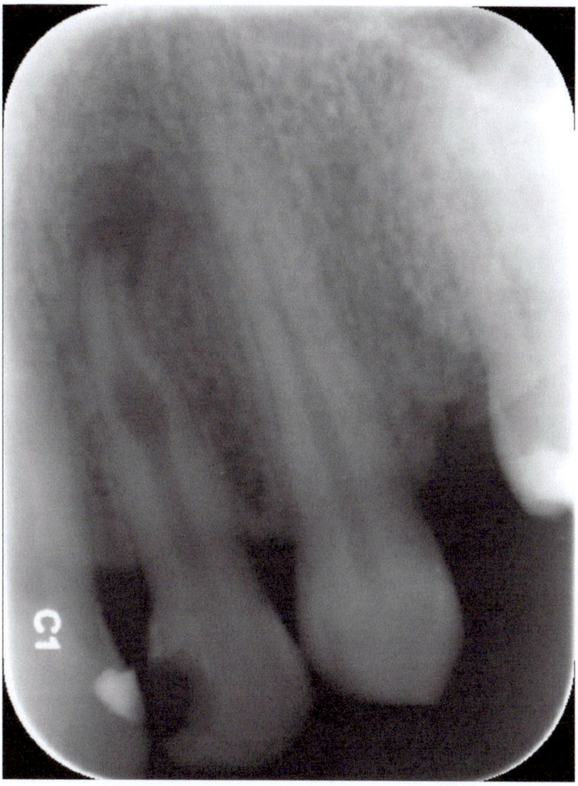

Fig. 20.12 Internal resorption (Courtesy of Dr. Jae Ha Jung)

affecting replanted avulsed teeth. Three types have been identified: surface, inflammatory, and replacement (Trope et al. 1992; Andreasen and Andreasen 1992).

(a) Surface Resorption

This type, also known as "repair-related resorption," is transient and shows as lacunae of resorption in the cementum. It is not usually visible on radiographs. Once the resorption is arrested, the lacunae are repaired by the deposition of new cementum. No treatment is usually indicated.

(b) Inflammatory (Infection-Related) Resorption

This type of resorption is a response to an infected root canal system in conjunction with injury to the periodontal ligament. It often occurs when the teeth are traumatized with the need for replantation, in addition to other types of luxation injuries. It is characterized by loss of tooth structure and adjacent alveolar bone. Inflammatory resorption is usually arrested after removal of the necrotic, infected pulp (Fig. 20.13).

(c) External Replacement (PDL-Related) Resorption

In this type of resorption, also known as ankylosis, the tooth structure is resorbed and replaced by bone. The common etiology is contusion of the PDL during intrusion, lateral luxation, or avulsion with subsequent replantation. When large areas of the PDL are traumatized, competitive wound healing processes begin, eventually resulting in replacement resorption. The

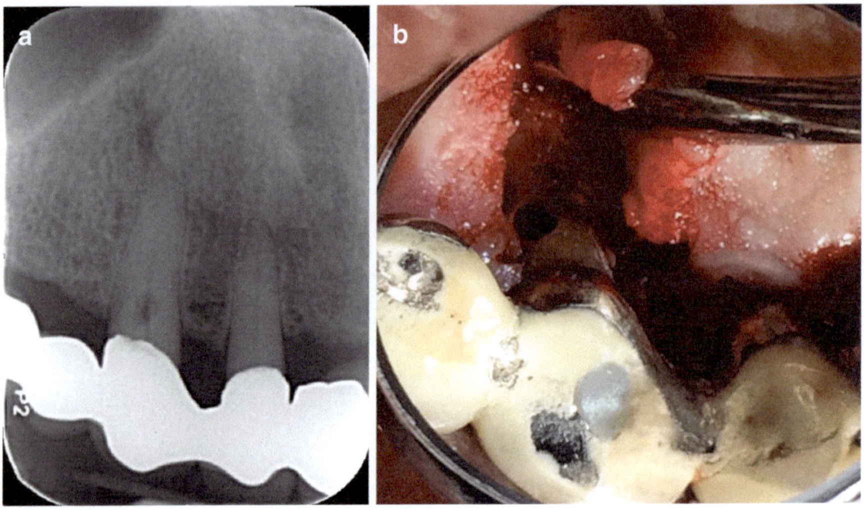

Fig. 20.13 External resorption. (**a**) Preoperative radiograph. (**b**) Surgical repair

characteristics of ankylosis are failure of the tooth to erupt along with adjacent teeth, lack of physiological mobility, and a "solid" metallic sound on percussion. Currently, no treatment to arrest replacement resorption has proved effective. It tends to continue until the root is replaced by bone.

20.3 Root Canal Anatomy

The root canal is the portion of the pulp cavity that extends from the canal orifice to the apical foramen. Its shape varies with the size, shape, and number of roots in different teeth. Root canals are commonly curved along their length and gradually sharpen. The root canal system is complex; canals can branch, divide, and rejoin. The number of root canals does not always correspond to the number of roots; roots that are elliptical or have a concave surface frequently have more than one canal. According to Orban (1957), the shape of the canal is largely determined by the shape of the root. Root canals can be round, tapering elliptical, broad, thin, etc., but it is important to know that a canal is seldom round at any level. This makes it hard clinically to clean and shape the root canal system with round-shaped instrument.

For good clinical practice and successful endodontic procedures, we need to know the normal root canal morphology and its frequent variations. Several studies have demonstrated these aspects of the root canal system. This knowledge helps clinicians during actual root canal procedures.

Weine (1996) categorized the canal systems in any root into four basic types. Using precise techniques, Vertucci (1984) established eight different pulp space configurations rather than four, which can be described briefly as follows:

- Type I: A single canal extends from the pulp chamber to the apex (one canal).
- Type II: Two separate canals leave the pulp chamber and join short of the apex to form one canal (coronal-apical, 2-1).
- Type III: One canal leaves the pulp chamber and divides into two in the root; the two then merge to exit as one canal (coronal-middle-apical, 1-2-1).
- Type IV: Two separate, distinct canals extend from the pulp chamber to the apex (two canals).
- Type V: One canal leaves the pulp chamber and divides short of the apex into two separate, distinct canals with separate apical foramina (coronal-apical, 1-2).
- Type VI: Two separate canals leave the pulp chamber, merge in the body of the root, and separate short of the apex to exit as two distinct canals (coronal-middle-apical, 2-1-2).
- Type VII: One canal leaves the pulp chamber, divides and then rejoins in the body of the root, and finally separates into two distinct canals short of the apex (pulp chamber-coronal-middle-apical, 1-2-1-2).
- Type VIII: Three separate, distinct canals extend from the pulp chamber to the apex (three canals) (Fig. 20.14).

Gender, racial, and ethnic data should be considered during preoperative evaluation, especially for a root canal procedure. Specific types of canal morphology occur in different racial groups.

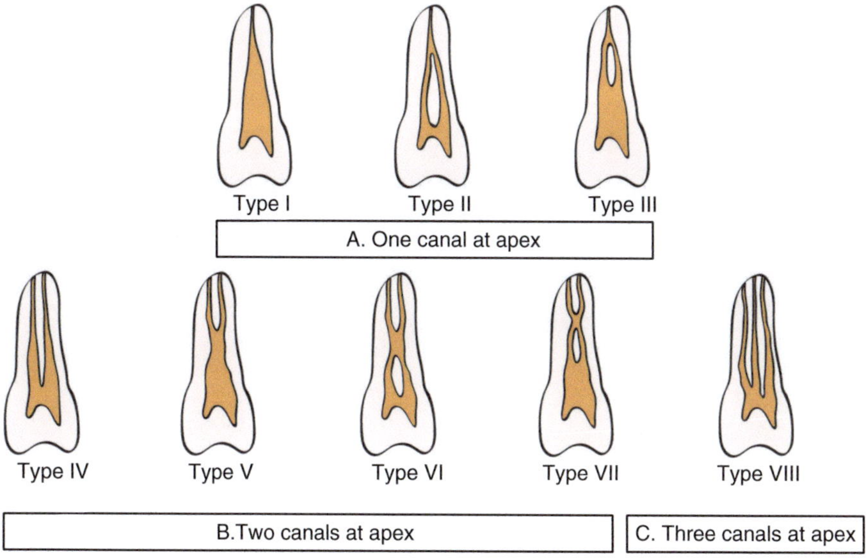

Fig. 20.14 Diagrammatic representation of canal configurations based on the work of Vertucci (1984). (**a**) One canal at apex. (**b**) Two canals at apex. (**c**) Three canals at apex

20.3.1 Other Variations

20.3.1.1 C-Shaped

A C-shaped canal is very common on the mandibular molars especially among Asians and those of Asian descent (Zheng et al. 2011). The main etiology of this anomaly is the failure of Hertwig's epithelial root sheath to fuse on either the buccal or lingual root surface. A C-shaped canal system can be observed in many variations in root and canal morphology. It can vary along the root depth so the appearances of the orifices are not necessarily good predictors of the actual canal anatomy. Fan et al. (2004) classified C-shaped canal configurations into five different types (Fig. 20.15):

- Category I (C1): The shape is an uninterrupted "C" with no separation or division.
- Category II (C2): The canal shape resembles a semicolon resulting from a discontinuity in the "C" outline, but either angle α or β should be no less than 60°.
- Category III (C3): Two or three separate canals and both angles, α and β, are less than 60°.
- Category IV (C4): Only one canal, which is round or oval in cross section.
- Category V (C5): No canal lumen can be observed (usually seen near the apex only).

The access cavity for teeth with a C-shaped root canal system varies considerably and depends on the pulp morphology of the specific tooth. Owing to the complex anatomy, teeth with C-shaped anatomy pose considerable technical difficulties. Patients with such teeth should be considered for referral to endodontists (Fig. 20.16).

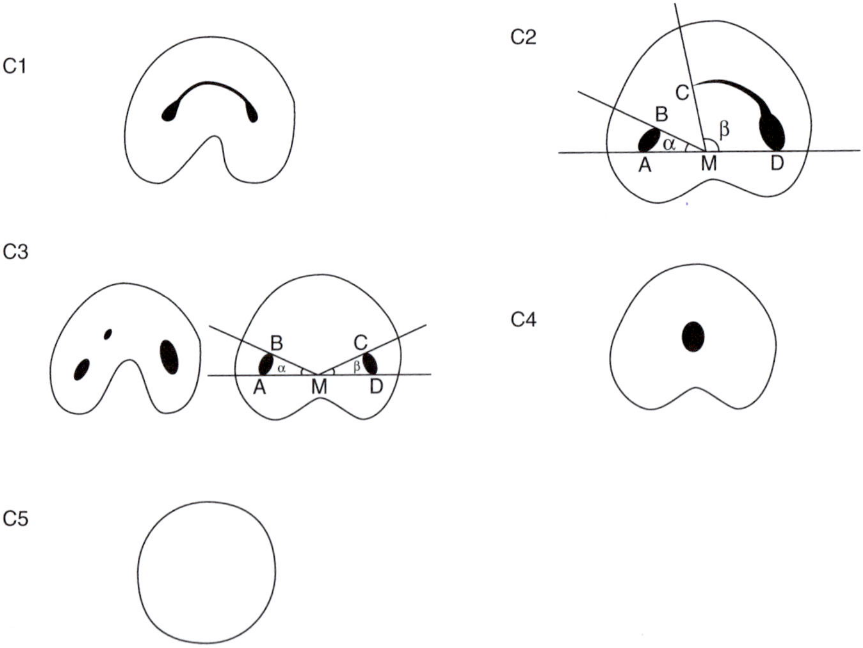

Fig. 20.15 Classification of C-shaped canal configuration (Fan et al. 2004)

20.3.1.2 Isthmus

Isthmus is defined as narrow, ribbon-shaped passage or anatomical communication between two root canals that contain pulp or pulpally derived tissue. It is commonly found in teeth with multiple roots. These structures can store various amounts of tissue, and the tissues often contain bacteria and their byproducts if the pulp is infected. Isthmus can be complete or incomplete. However, the isthmus should ideally be cleaned, shaped, and obturated.

Kim and colleagues offered a classification of isthmus (Kim et al. 2001). Type I is an incomplete isthmus; it is a faint communication between two canals. Type II is characterized by two canals with a definite connection between them (complete isthmus). Type III is a short, complete isthmus between two canals. Type IV is a complete or incomplete isthmus among three or more canals. Type V is marked by two or three canal openings without visible connections (Fig. 20.17).

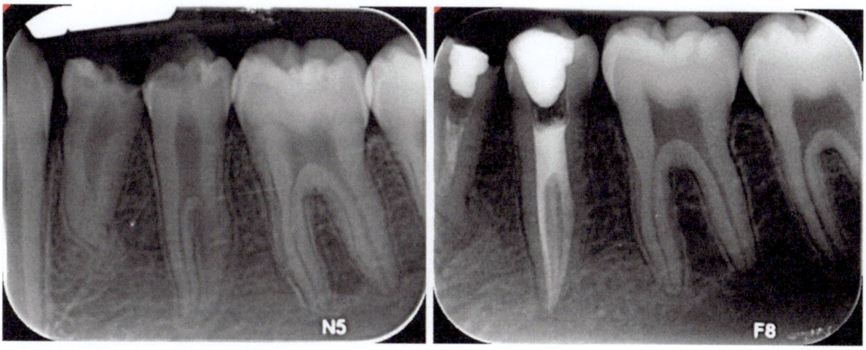

Fig. 20.16 C-shaped root canal #20

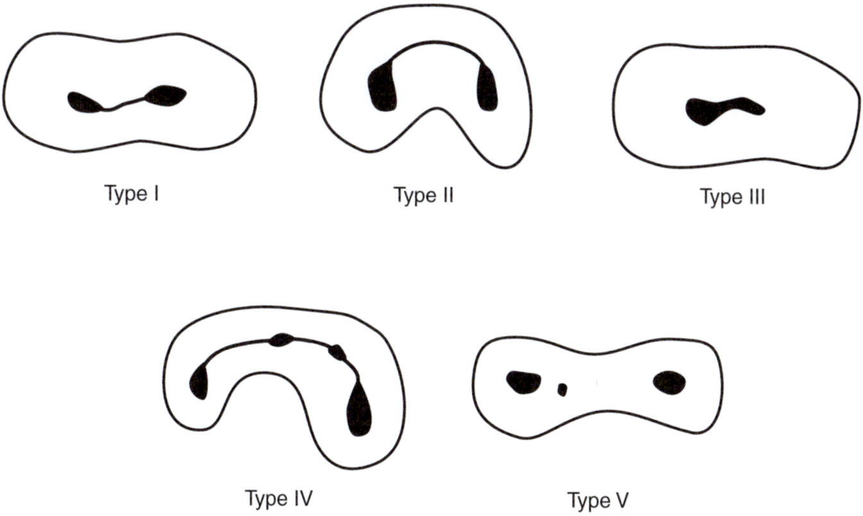

Fig. 20.17 Schematic representation of isthmus classifications (Kim et al. 2001)

New technologies such as CBCT can provide three-dimensional morphology information concerning many more irregular and challenging root canal anatomies (Peters et al. 2000).

References

Altug-Atac AT, Erdem D (2007) Prevalence and distribution of dental anomalies in orthodontic patients. Am J Orthod Dentofac Orthop 131:510–514

Andreasen JO, Andreasen FM (1992) Root resorption following traumatic dental injuries. Proc Finn Dent Soc 88:95–114

Bäckman B, Holm AK (1986) Amelogenesis imperfecta: prevalence and incidence in a northern Swedish county. Community Dent Oral Epidemiol 14:43–47

Ball RL, Barbizam JV, Cohenca N (2013) Intraoperative endodontic applications of cone beam computed tomography. J Endod 39:548

Chosack A, Eidelman E, Wisotski I et al (1979) Amelogenesis imperfecta among Israeli Jews and the description of a new type of local hypoplastic autosomal recessive amelogenesis imperfecta. Oral Surg Oral Med Oral Pathol 47:148–156

Dexton AJ, Arundas D, Rameshkumar M et al (2011) Retreatodontics in maxillary lateral incisor with supernumerary root. Conserv Dent 14:322–324

Fan B, Cheung G, Fan M et al (2004) C-shaped canal system in mandibular second molars. I. Anatomical fractures. J Endod 30:899

Järvinen S, Lehtinen L, Milén A (1980) Epidemiologic study of joined primary teeth in Finnish children. Community Dent Oral Epidemiol 8:201–202

Kim S, Pecora G, Rubinstein R et al (2001) Color atlas of microsurgery in endodontics. Saunders, Philadelphia

Kuyk JK, Walton RE (1990) Comparison of the radiographic appearance of root canal size to its actual diameter. J Endod 16:528

Law L, Fishelberg G, Skribner JE et al (1994) Endodontic treatment of mandibular molars with concrescence. J Endod 20:562–564

Lee CK, King NM, Lo EC et al (2006) Talon cusp in the primary dentition:literature review and report of three rare cases. J Clin Pediatr Dent 30:299–305

Matherne RP, Angelopoulos C, Kulild JC et al (2008) Use of cone-beam computed tomography to identify root canal systems in vitro. J Endod 34:87–89

Mohan B (2014) Hypercementosis and concrescence of maxillary second molar with third molar: a case report and review of literature. Oral Health Dent Manag 13:558–561

Moskow BS, Canut PM (1990) Studies on root enamel (2). Enamel pearls. A review of their morphology, localization, nomenclature, occurrence, classification, histogenesis and incidence. J Clin Periodontol 17:275–281

Neville BW, Damm DD, Allen CM et al (1993) Abnormalities of teeth. In: Oral and maxillofacial pathology, 3rd edn. Saunders, Philadelphia, pp 54–119

O Carroll MK (1990) Fusion and gemination in alternate dentitions. Oral Surg Oral Med Oral Pathol 69:655

Oehlers FA (1957) Dens invaginatus I. Variations of the invagination process and associated anterior crown forms. Oral Surg Oral Med Oral Pathol 10:1204–1218

Orban BJ (1957) Oral histology and embryology, 4th edn. Mosby, St. Louis

Peters OA, Laib A, Rüegsegger P et al (2000) Three-dimensional analysis of root canal geometry by high resolution computed tomography. J Dent Res 79:1405

Sedano HO (1975) Congenital oral anomalies in Argentinian children. Community Dent Oral Epidemiol 3:61–63

Seow WK, Taji S (2011) Diagnosis and management of toothwear in children. In: Khan F, Young WG (eds) Toothwear: the ABC of the worn dentition. Wiley-Blackwell, Chichester, pp 16–33

Sundqvist G, Lerner UH (1996) Bradykinin and thrombin synergistically potentiate interleukin 1 and tumour necrosis factor induced prostanoid biosynthesis in human dental pulp fibroblasts. Cytokine 8:168

Swan RH, Hurt WC (1976) Cervical enamel projections as an etiologic factor in furcation involvement. J Am Dent Assoc 93:342–345

Trope M, Yesilsoy C, Koren L et al (1992) Effect of different endodontic treatment protocols on periodontal repair and root resorption of replanted dog teeth. J Endod 18:492–496

Vertucci FJ (1984) Root canal anatomy of the human permanent teeth. Oral Surg Oral Med Oral Pathol 58:589

Walton RE, Leonard LA (1986) Cracked tooth: an etiology for "idiopathic" internal resorption? J Endod 12:167

Weine FS (ed) (1996) Endodontic therapy, 5th edn, Mosby, St. Louis, p 243

Witkop CJ Jr (1971) Manifestations of genetic diseases in the human pulp. Oral Surg Oral Med Oral Pathol 32:278–316

Yusof WZ (1990) Non-syndromal multiple supernumerary teeth: literature review. J Can Dent Assoc 56:147–149

Zheng Q, Zhang L, Zhou X et al (2011) C-shaped root canal system in mandibular second molars in a Chinese population evaluated by cone-beam computed tomography. Int Endod J 44:857

Abnormal Tooth Position

21

Masayoshi Uezono and Keiji Moriyama

21.1 Introduction

This chapter reviews impacted teeth, rotated teeth, transposition, and ectopic eruption, which are classified as major abnormal tooth positions in clinical dentistry.

21.2 Impacted Teeth

An impacted tooth can be defined as one that does not erupt into its correct position in the dental arch within the expected time frame and instead remains below the gingival line (Hattab and Abu Alhaija 1999) resulting in lost eruption potential. Alternatively, impacted teeth have been defined as those that are prevented from erupting into their proper positions by a physical barrier along their eruption path (Farman 2007). With regard to tooth type, it has been reported that the frequency of impaction is highest for the third maxillary molars, followed by the third mandibular molars. Although it is less frequent than for the third molars, a relatively high rate of impaction has also been reported for the maxillary canines (Dachi and Howell 1961; Kramer and Williams 1970). No differences between sexes in the rates of impaction of the third molars have been reported, but in one study, maxillary cuspid impactions were detected more frequently in female patients than in male patients (Dachi and Howell 1961).

M. Uezono (✉) · K. Moriyama
Department of Maxillofacial Orthognathics, Graduate School of Tokyo Medical and Dental University, Tokyo, Japan
e-mail: uezomort@tmd.ac.jp

J. Iwanaga, R. S. Tubbs (eds.), *Anatomical Variations in Clinical Dentistry*,
https://doi.org/10.1007/978-3-319-97961-8_21

21.2.1 Etiology

Both general and local factors are reportedly associated with tooth impaction. General factors include cleidocranial dysplasia, osteopetrosis, rickets, and abnormal parathyroid hormone secretion (Almonaitiene et al. 2010). Local factors include obstruction of tooth eruption, insufficient space for eruption, ectopic position of the tooth bud, a history of trauma to the primary teeth, ankylosis, hyperdontia, odontomas, and dentigerous cysts (Otsuka et al. 2001). There are racial differences in the frequency of tooth impaction, especially for the maxillary canines. Tooth impaction also exhibits a familial phenotypic expression indicating a possible genetic etiology, although specific details in this regard remain unclear. Furthermore, because the distal surface of the lateral incisors functions as a guide for the eruption of the canines, the congenital absence (or microdontia) of a lateral incisor can also cause impaction (Litsas and Acar 2011; Becker and Chaushu 2015).

21.2.2 Classification

The following classifications of impaction have been reported for the maxillary and mandibular third molars and maxillary canines.

21.2.2.1 Classification of Impacted Mandibular Third Molars

Winter (1926) classified impacted mandibular third molars by their angulation. Winter also developed a system for assessing the ease or difficulty of extracting mandibular third molars, called Winter's lines (Fig. 21.1). Winter's lines include the white line, which indicates the occlusal plane and the difference in occlusal level of the second and third molars; the amber line, which indicates bone level and denotes the alveolar bone covering the impacted tooth and the portion of tooth not covered by the bone; and the red line, which represents depth and indicates the vertical position of impacted tooth within the mandible. The red line is also important during tooth extraction as an indicator of how much bone has to be removed in the process. For example, tooth extraction using local anesthesia is possible when the red line is less than 5 mm, but when it is greater than 5 mm the use of a general anesthetic or intravenous sedation is recommended. However, Kumar et al. (2014) concluded that Winter's "red line" is not reliable and of little practical value in assessing the difficulty or depth in the removal of an impacted wisdom tooth.

Pell and Gregory (1933) classified the position of the impacted mandibular third molars based on the relationship between the impacted third molar and the ramus of the mandible or the second molar (Fig. 21.2). In Pell-Gregory's classification, the horizontal position (class I–III) is determined by the space between the distal aspect of the second molar and the anterior border of the ramus, and the vertical position (position A–C) is determined by the occlusal plane of the second molars and the cervical margin of the second molars. As class and position increase, the degree of difficulty of tooth extraction increases, so it is used as an index for surgical procedure selection. Pederson also proposed adding the angulation of the impacted tooth

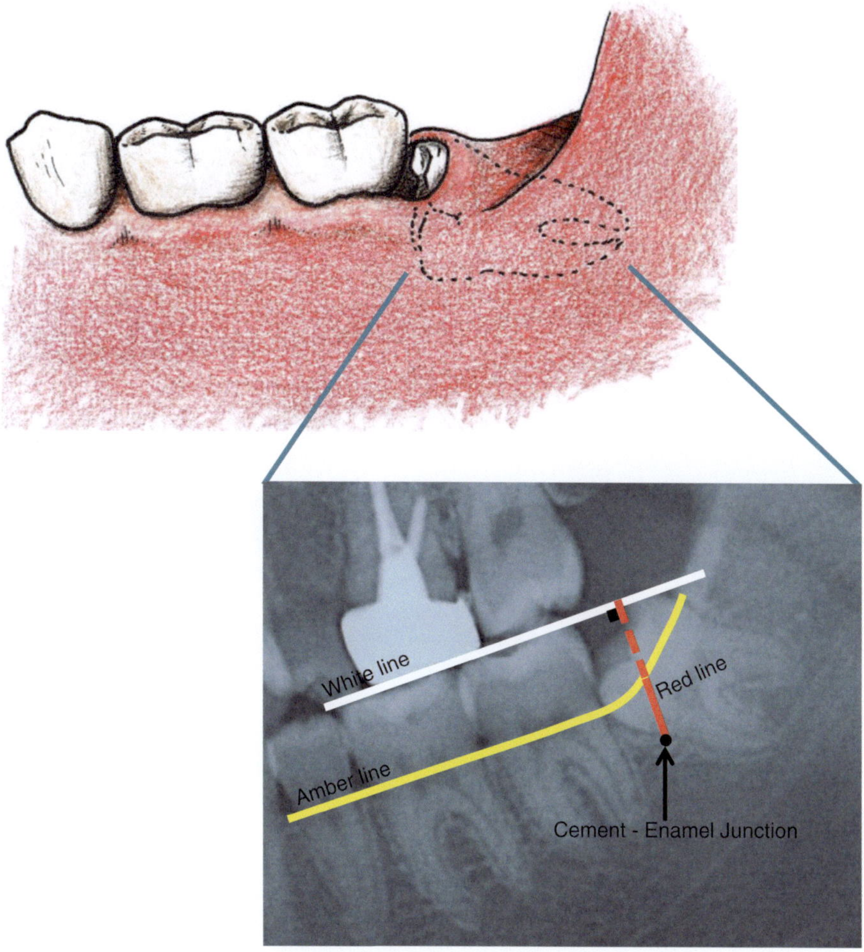

Fig. 21.1 Winter's line

to the Pell-Gregory's classification to enhance the reliability of assessing the difficulty of impacted tooth extraction (Pederson 1988).

21.2.2.2 Classification of Impacted Maxillary Third Molars

Winter (Hashemipour et al. 2013) and Archer (1975) have classified impacted teeth by angulation, as was done for the mandibular third molars.

21.2.2.3 Classification of Impacted Maxillary Canines

Archer (1975), Field, and Ackerman (Laskin 1985) have classified impacted maxillary canines according to the buccolingual position, axial direction of the tooth, and position in relation to the roots of the adjacent teeth. Due to the importance of maxillary canines with respect to both esthetics and function, it is desirable to induce

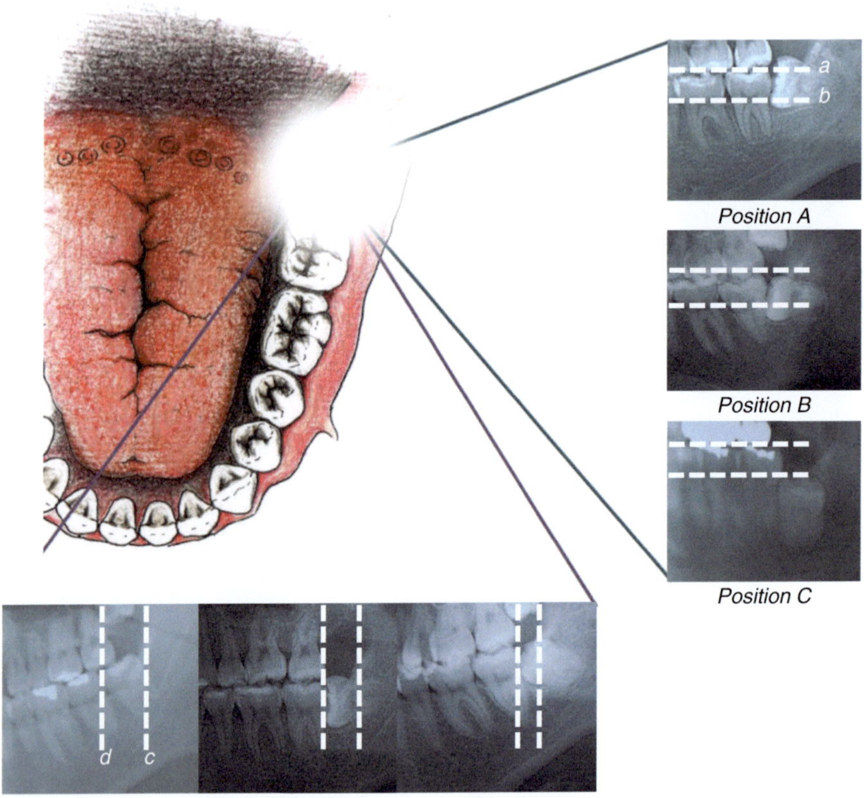

Fig. 21.2 Pell-Gregory's classification. Position A. The highest portion of impacted third molar is on a level with or above the occlusal plane of the second molar (**a**). Position B. The highest portion of impacted third molar is below the occlusal plane but above the cervical line of the second molar (**b**). Position C. The highest portion of impacted third molar is below the cervical line of the second molar. Class I. There is sufficient space available between the anterior border of the ascending ramus (**c**) and the distal aspect of the second molar (**d**) for the eruption of the third molar. Class II. The space available between the anterior border of the ramus and the distal aspect of the second molar is less than the mesiodistal width of the crown of the third molar. It denotes that the distal portion of the third molar crown is covered by bone of the ascending ramus. Class III. The third molar is totally embedded in the bone of the anterior border of the ascending ramus because of the absolute lack of space

eruption without extraction. As a way of evaluating the danger of root resorption around the impacted tooth and the requirement for fenestration and traction, Ericson and Kurol (1988, 1987) have proposed a system involving sector classification based on the extent of overlap between the apices of the canines and the roots of the central and lateral incisors, as well as a method for measuring the vertical distance against the angle of mesial inclination and the virtual occlusal plane (Fig. 21.3). They used angle α to represent the angle formed between the interincisor midline

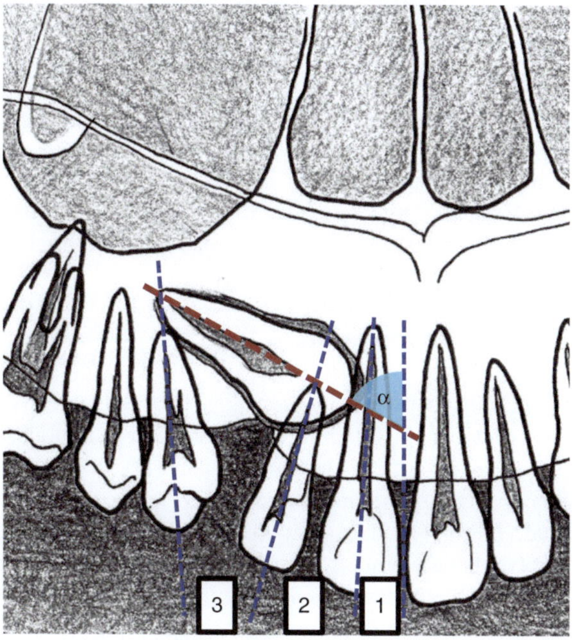

Fig. 21.3 Sector classification by Ericson and Kurol. Sector 1: The cusp tip of the canine is between the interincisor median line and the long axis of the central incisor. Sector 2: The peak of the cuspid of the canine is between the major axes of the lateral and central incisors. Sector 3: The peak of the cuspid of the canine is between the major axis of the lateral and the first premolar

and long axis of the canine. The risk of resorption of the root of the lateral incisor increases by 50% if the cusp of the canine belongs to sector 1 or 2 and if α angle is greater than 25°. The duration of treatment is longer if the canine is found in sector 1 and shorter if it belongs to sector 3, with respect to sector 2. The necessity of treatment and the degree of treatment difficulty increases as this angle increases.

21.3 Rotated Teeth

Rotation refers to movement that occurs around a tooth's long axis. Tooth rotation occurs most frequently with the cuspids, followed by the second bicuspids, lateral incisors, central incisors, and first molars (in this order (Swanson et al. 1975)). Winging of the central incisors is a typical example, and in severe cases, rotation of up to 180° is observed.

21.3.1 Etiology

Causes of tooth rotation include severe crowding, supernumerary teeth, odontomas, alveolar clefts (Ranta 1986), over-retained primary teeth, ectopic canines, and scar tissue from trauma. External forces exerted by soft tissues such as masticatory forces and forces from the tongue can also contribute to tooth rotation, as can habits such as digit sucking (Köhler and Holst 1973), lip biting, and tongue thrusting.

21.3.2 Classification

Tooth rotation is classified as either pure rotation or generalized rotation. Pure rotation is subclassified into transverse and long-axis rotation. Transverse rotation includes tipping and torqueing, while long-axis rotation is defined as the rotation of a tooth around its long axis. Generalized rotation is any movement that is not pure translation or pure rotation and is seen in day-to-day practice. It is classified by rotation direction as mesiolingual, distobuccal, distolingual, or mesiobuccal.

21.4 Transposition

Tooth transposition is the positional interchange of two adjacent teeth, especially their roots, or the development or eruption of a tooth in a position normally occupied by a non-adjacent tooth (Peck et al. 1993). It is a rare and abnormal tooth position and is more commonly observed in the maxilla than the mandible (Yoshida et al. 1995). The transposition of teeth usually involves the canines but can also involve the lateral incisors and premolars. Canine-lateral incisor transpositions are reportedly more common than canine-premolar transpositions, and it has also been reported that most transpositions are bilateral involving the same pairs. In unilateral cases, there is a predominance of left-sided transposition.

21.4.1 Etiology

The causes of tooth transposition include irregularities in tooth bud position, deviation in the direction of eruption, the presence of supernumerary teeth, early loss or late retention of primary teeth, and a history of dental trauma. Genetic causes have been reported, but due to the low incidence rate, the details of this aspect of tooth transposition remain unclear (Kaneko et al. 2001).

21.4.2 Classification

The most important factor in the treatment of transposition is determining the position where eruption should be induced. To aid in deciding whether to induce eruption in the original position or in the position where the tooth has moved to, Shapira and Kuftinec (1989) used panoramic X-ray imaging to classify tooth transposition as "complete" or "incomplete" according to the position of the canine root apex and the inclination of the tooth axis. Transposition is "complete" if both the crown and the root apex of the tooth have moved and "incomplete" if the crown has moved but the root apex is still in the normal position. The position of the root apex is an important factor when determining the position of eruption, and if the position of the root apex has completely relocated, it is necessary to consider an arrangement and order that deviates from the norm.

21.5 Ectopic Eruption

Before the permanent teeth erupt into the oral cavity and become visible, they move through the bone along their eruption path. When the eruption path is incorrect, the tooth will erupt in the mouth in an incorrect position or may not erupt at all. Abnormal eruption at an undesirable location is called "ectopic eruption" (Yaseen et al. 2011). Maxillary first molars exhibit the highest rate of ectopic eruptions, and while extremely rare, cases of ectopic eruption of maxillary canines and lateral incisors have been reported, as have the eruption of supernumerary teeth into the nasal cavity (Yamasaki and Fujimoto 1982).

21.5.1 Etiology

Meara and Williams (1962) reported insufficient width between canines and insufficient longitudinal growth of the maxilla and mandible as main causes of ectopic eruption. Chapman (1923) has suggested that the main causes of ectopic eruption of maxillary first molars are small arches, deviant paths of eruption of the permanent molar, lack of forward movement of all primary teeth, and early eruption of the maxillary first permanent molars. Pulver (1968) investigated tooth sizes and features of maxillary dental arches in cases of ectopic eruption of maxillary first molars. The average size of the maxillary primary and permanent teeth was significantly greater than that of the control group. Furthermore, in unilateral cases the first molars and second primary molars of the affected side were significantly larger than those of the unaffected side. With regard to maxillary dental arches, it was reported that the occipitofrontal diameter was significantly smaller than that of the known standards and that the maxilla was positioned farther than usual from the skull base.

21.5.2 Classification

Sweet (1939) devised the following ectopic eruption classifications: eruption of mandibular permanent lateral incisors initiating the loss of the primary canines, eruption of maxillary first permanent molars initiating the loss of second primary molars, eruption of maxillary permanent lateral incisors initiating the loss of primary canines, and eruption of mandibular first permanent molars initiating loss of primary second molars.

With regard to ectopic eruption of maxillary first molars—the tooth type that exhibits the highest frequency of ectopic eruption—Young (1957) classified the eruptions as either "jump type" or "hold type" according to the relationship with the maxillary second primary molar. In the jump type, the first molar erupts distal of the second primary molar, despite the first molar absorbing the distal root of the primary second molar. Conversely, in the hold type early loss of the second primary molar is observed because most of the first molar is impacted beneath the second primary molar. The hold type is observed frequently and comprises 30–50% of ectopic eruptions. Early loss of the second primary molar and abnormal eruption position of the first molar causes additional dentition and occlusion abnormality. Thus, the early detection and early treatment of ectopic eruption are important.

References

Almonaitiene R, Balciuniene I, Tutkuviene J (2010) Factors influencing permanent teeth eruption. Part one—general factors. Stomatologija 12:67–72

Archer HW (1975) Oral and maxillofacial surgery (ed 5). WB Saunders, Philadelphia, PA

Becker A, Chaushu S (2015) Etiology of maxillary canine impaction: a review. Am J Orthod Dentofac Orthop 148:557–567

Chapman H (1923) First upper permanent molar partially impacted against second deciduous molar. Int J Ortho Oral Surg Radiol 9:339–345

Dachi SF, Howell FV (1961) A survey of 3,874 routine full-mouth radiographs. I. A study of retained roots and teeth. Oral Surg Oral Med Oral Pathol 14:916–924

Ericson S, Kurol J (1987) Radiographic examination of ectopically erupting maxillary canines. Am J Orthod Dentofac Orthop 91:483–492

Ericson S, Kurol J (1988) Early treatment of palatally erupting maxillary canines by extraction of the primary canines. Eur J Orthod 10:283–295

Farman AG (2007) Tooth eruption and dental impaction. Panoramic radiology: seminars on maxillofacial imaging and interpretation. Springer, New York, pp 73–82

Hashemipour MA, Tahmasbi-Arashlow M, Fahimi-Hanzaei F (2013) Incidence of impacted mandibular and maxillary third molars: a radiographic study in a Southeast Iran population. Med Oral Patol Oral Cir Bucal 18:e140–e145

Hattab FN, Abu Alhaija ES (1999) Radiographic evaluation of mandibular third molar eruption. Oral Surg Oral Med Oral Path Oral Radiol Endod 88:285

Kaneko T, Okamoto T, Koshikawa M et al (2001) Consideration of migrated tooth in a family. J Hokkaido Orthod Soc 29:3–10

Köhler L, Holst K (1973) Malocclusion and sucking habits of four-year-old children. Acta Paediatr Scand 62:373–379

Kramer RM, Williams AC (1970) The incidence of impacted teeth. A survey at Harlem hospital. Oral Surg Oral Med Oral Pathol 29:237–241

Kumar S, Reddy MP, Chandra L et al (2014) The "red line" conundrum: a concept beyond its expiry date? J Maxillofac Oral Surg 13:612–614

Laskin DM (ed) (1985) Oral and maxillofacial surgery: oral surgery. CV Mosby, St. Louis (MO)

Litsas G, Acar A (2011) A review of early displaced maxillary canines: etiology, diagnosis and interceptive treatment. Open Dent J 16:39–47

Meara O, Williams F (1962) Ectopic eruption pattern in selected permanent teeth. J Dent Res 41:607–616

Otsuka Y, Mitomi T, Tomizawa M et al (2001) A review of clinical features in 13 cases of impacted primary teeth. Int J Paediatr Dent 11:57–63

Peck L, Peck S, Attia Y (1993) Maxillary canine-first premolar transposition, associated dental anomalies and genetic basis. Angle Orthod 63:99–109. discussion 110

Pederson GW (1988) Oral surgery. WB Saunders, Philadelphia. (Cited in: Koerner KR. The removal of impacted third molars—principles and procedures. Dent Clin North Am 1994;38:255–78

Pell GJ, Gregory BT (1933) Impacted mandibular third molars: classification and modified techniques for removal. Dent Digest 39:330–338

Pulver F (1968) Etiology and prevalence of ectopic eruption of maxillary first permanent molar. J Dent Child 35:138–146

Ranta R (1986) A review of tooth formation in children with cleft lip/palate. Am J Orthod Dentofac Orthop 90:11–18

Shapira Y, Kuftinec MM (1989) Tooth transposition - review of the literature and treatment considerations. Angle Orthod 59:271–276

Swanson WD, Riedel RA, D'Anna JA (1975) Postretention study: incidence and stability of rotated teeth in humans. Angle Orthod 45:198–203

Sweet CA (1939) Ectopic eruption of permanent tooth. J Am Dent Assoc 26:574–579

Winter GB (1926) Impacted mandibular third molar. American Medical Book Co, St. Louis

Yamasaki Y, Fujimoto M (1982) An inverted tooth in the nasal cavity. Jibi to Rinsho 28:589–592

Yaseen SM, Naik S, Uloopi KS (2011) Ectopic eruption - a review and case report. Contemp Clin Dent 2:3–7

Yoshida S, Suzuki A, Ito K et al (1995) Current status of migrated tooth and clinical consideration for the orthodontic treatment. J Hiroshima Univ Dent Soc 27:266–274

Young DH (1957) Ectopic eruption of permanent first molar. J Dent Child 24:153–162

MIX
Papier aus verantwortungsvollen Quellen
Paper from responsible sources
FSC® C105338

If you have any concerns about our products,
you can contact us on
ProductSafety@springernature.com

In case Publisher is established outside the EU,
the EU authorized representative is:
Springer Nature Customer Service Center GmbH
Europaplatz 3, 69115 Heidelberg, Germany

Printed by Libri Plureos GmbH
in Hamburg, Germany